LECTURE NOTES ON IMPEDANCE SPECTROSCOPY

Lecture Notes on Impedance Spectroscopy

Measurement, Modeling and Applications

Editor

Olfa Kanoun

Chair for Measurement and Sensor Technology
Chemnitz University of Technology, Chemnitz, Germany

VOLUME 2

CRC Press
Taylor & Francis Group
Boca Raton London New York

CRC Press is an imprint of the
Taylor & Francis Group, an **informa** business

A BALKEMA BOOK

CRC Press
Taylor & Francis Group
6000 Broken Sound Parkway NW, Suite 300
Boca Raton, FL 33487-2742

First issued in paperback 2018

Typeset by Vikatan Publishing Solutions (P) Ltd., Chennai, India

ISBN-13: 978-0-415-69838-2 (hbk)
ISBN-13: 978-1-138-11195-0 (pbk)

**Visit the Taylor & Francis Web site at
http://www.taylorandfrancis.com**

**and the CRC Press Web site at
http://www.crcpress.com**

Lecture Notes on Impedance Spectroscopy – Kanoun (ed)
© *2012 Taylor & Francis Group, London, ISBN 978-0-415-69838-2*

Table of contents

Lecture Notes on Impedance Spectroscopy – Kanoun (ed)
© *2012 Taylor & Francis Group, London, ISBN 978-0-415-69838-2*

Preface

Impedance Spectroscopy is a powerful measurement method used in many application fields such as electrochemistry, material science, biology and medicine, semiconductor industry and sensors.

Using the complex impedance at various frequencies increases the informational basis that can be gained during a measurement. It helps to separate different effects that contribute to a measurement and, together with advanced mathematical methods, non-accessible quantities can be calculated.

Dealing with Impedance Spectroscopy in general requires competences in several fields of research, such as measurement technology, electrochemistry, modeling, mathematical and physical methods and nonlinear optimization. Depending on the specific challenges of the considered application there are generally more efforts to be done in one or two specific fields. The scientific dialogue between specialists of Impedance Spectroscopy, working with different applications, is therefore particularly profitable and inspiring.

The International Workshop on Impedance Spectroscopy has been launched already in June 2008 with the aim to serve as a platform for specialists and users to share experiences with each other. Since 2009 it became an international workshop addressing more fundamentals and application fields of impedance spectroscopy. The workshop is gaining increasingly more acceptance in the scientific and industrial fields.

This book is the second in the series Lecture Notes on Impedance Spectroscopy. It includes selected and extended contributions from the International Workshop on Impedance Spectroscopy (IWIS'10). It is a set of presented contributions of world-class manuscripts describing state-of-the-art research in the field of impedance spectroscopy. It reports about new advances and different approaches in dealing with impedance spectroscopy including theory, methods and applications. The book is interesting for research and development in the field of impedance spectroscopy.

I thank all contributors for the interesting contributions, for confidence and for having patience with us during the preparation of the proceedings.

Prof. Dr.-Ing. Olfa Kanoun

Lecture Notes on Impedance Spectroscopy – Kanoun (ed)
© *2012 Taylor & Francis Group, London, ISBN 978-0-415-69838-2*

Modeling impedance spectroscopy data from lithium-ion battery aging utilizing particle swarm optimization

Joachim Georg Roth, Meike Slocinski & Johann-Friedrich Luy
Daimler AG, Department for Quality Analysis, Ulm, Germany

ABSTRACT: The increasing number of hybrid electrical vehicles (HEV) and electrical vehicles (EV) on the market requires new and reliable energy storage systems. Lithium-ion batteries are the dominating technology nowadays. Lifetime prediction needs precise measurement methods and robust indicators for determining the present state of health (SOH). For predicting remaining lifetime, detailed knowledge of the aging process is required.

In this work we present results of aging experiments where automotive li-ion high power cells were exposed to high temperatures or cyclic charging and discharging. During the aging process the state of the cells was repeatedly measured, not only with conventional pulse and discharge tests but also with impedance spectroscopy. This promising measurement method allows a detailed insight into the cell's complex impedance in the frequency range yielding additional information that enables more detailed analyses compared to current pulse tests. Measured curves were modeled by an underlying equivalent circuit model containing constant phase elements which fit the shape characteristics of the impedance spectra much more accurately than RC circuits. The fitting algorithm utilizes particle swarm optimization for a higher precision in curve approximation. Stable aging indicators were extracted leading to a detailed description of the aging process using a small number of parameters.

Keywords: Impedance Spectroscopy, Battery Aging, Battery Testing, Modeling, Particle Swarm Optimization

1 MOTIVATION

Due to the dwindling of fossil energy resources and the need for reduction of combustion products, the electrification of power trains is the main long-term goal of the automobile industry. Nowadays, hybrid electrical vehicles (HEV) are a transitional solution. In the future the full electrification will be the key to achieve locally emission-free mobility. For all kinds of HEV and electrical vehicles (EV) the energy storage system is the key for a good usability. From a car manufacturer's viewpoint a high quality of the product is directly related to the quality of the battery.

Nowadays lithium-ion batteries are the technology of choice. One big challenge is the precise knowledge of the current state of health (SOH) of the li-ion battery (defined by the SOH of each single cell) under use. Therefore, detailed analyses of the aging behavior in the development stage are a prerequisite. Even in the lab reliable SOH measurements are a challenge, since these tests require long preparation and measurement times in the range of hours. This is why we use an alternative measurement method to the common pulse and discharge test for determining state indicators. During the last years electrochemical impedance spectroscopy (EIS) has been established as a promising measurement method for state determination of li-ion cells. The main challenge is the interpretation of the impedance spectra because of their high dimensionality compared to scalar parameters like common aging indicators i.e. inner resistance or capacity. Here we will present the process to extract the information from the measured impedance spectra by extracting meaningful scalar parameters out of the measured vectorial data. This dimension reduction is done by adjusting the parameters of an underlying model, which is in this case an electrical equivalent circuit containing lumped and distributed elements. The fitting of these parameters is accomplished by a non-deterministic algorithm, the Particle Swarm Optimization (PSO). This algorithm is inspired by nature and uses the advantages of swarm intelligence for fitting the model parameters to the measured impedance spectra. In this work we present two typical aging experiments on high power li-ion cells under lab conditions. The first experiment simulates a strongly accelerated calendar aging, the second experiment addresses aging due to cycle stress. The main purpose of both experiments is to discover the ongoing aging process of the cells. On this data we will compare the state indicators of common pulse and discharge tests with the results from impedance spectroscopy. Meaningful state indicators from EIS measurements will be extracted from the fitted parameters of the electrical equivalent circuit model. We will conclude that EIS measurements have the potential to replace common pulse and discharge tests.

2 BATTERY AGING AND MEASUREMENT METHODS

The performance of li-ion batteries worsens with calendar age and due to usage (M. Broussely and Staniewicz 2005), (G. Sarre and Broussely 2004). Aging experiments in the lab usually look at these two influences separately. The experiment investigating calendar aging is carried out at elevated temperatures with the influences temperature and state of charge (SOC) to adjust. The second experiment simulates usage of the battery. Here a lot of parameters can be varied: charge and discharge currents, cycle depth, temperatures and relaxation times are the most important ones, but this list is probably not complete.

During the aging experiments cells were measured repeatedly to detect even small changes in parameters. Depending on the acceleration of the aging experiments degradation may proceed very slowly. This requires highly precise measurement methods. A common method for state determination is the pulse and discharge test, which yields inner resistances and the capacity of the cells (Jossen and Weydanz 2006). Here the inner resistances do not represent a physical resistor, it is a time dependent parameter describing the voltage drop of a battery when a certain current is applied. Impedance spectroscopy is an alternative method acting in the frequency domain, where the complex impedance of the cells is determined (Macdonald 2005). The result is the impedance spectrum which can be displayed in the Gaussian plane or in a Bode plot. The modeling process of the EIS data extracts several state indicators, some of them will be identified as aging indicators and will be tracked over age.

3 EXPERIMENTAL

This section describes the two aging experiments. Both were carried out simultaneously on three Saft VL6P high power round cells with a nominal capacity of $7Ah$ until the end of life criterion of 80% of the initial capacity (Jossen and Weydanz 2006), (G. Sarre and Broussely 2004) was reached at the end of the experiments. The parameters shown are plotted only for one cell. Both experiments reflect a strongly accelerated aging compared to the standard usage in an HEV. Due to this acceleration it is very difficult to make predictions about aging behavior under normal conditions, but the experiments provide valuable information about the degradation of li-ion cells with ongoing age.

3.1 *Measurements for monitoring changes in SOH*

As described in section 2 we investigated the cells repeatedly, both with conventional pulse and discharge test and impedance spectroscopy in order to determine changes due to degradation. All measurements were carried out at $25°C$. For the pulse and discharge test the aging indicators are the inner resistance at 100% SOC and 5s after a current pulse of $-40A$ was applied, and the capacity determined by a one hour discharge. Impedance spectroscopy measurements were carried out with a self developed EIS measurement setup at 100% SOC. The detailed EIS measurement parameterization is given in table 1.

The procedure for determining state indicators from pulse and discharge tests as well as from EIS measurements is described in the flow chart figure 1. The modeling of impedance spectra is discussed more detailed in section 4.

3.2 Calendar aging experiment

The calendar aging experiment was carried out in a temperature chamber at $85°C$ with constant SOC of 100%. Due to the high temperature and high SOC the end of life criterion was reached within one month. Figures 2 and 3 show the development of the common state indicators inner resistance and capacity. Figure 4 shows the measured impedance spectra at different points of age. The span of the spectra increases and the characteristic shape changes during the aging process.

3.3 Cyclic aging experiment

The cyclic aging experiment was conducted with high currents of $\pm70A$ ($10C$ rate) and with a cycle depth of 20% SOC. Even though $\pm70A$ are high currents the specifications of these high power cells even allow higher C rates. The experiment followed a weekly procedure, namely four days of cycling without any relaxation followed by three days of relaxation time. The tests for state determination were performed during the relaxation periods. Figures 5 and 6 show the trends of inner resistance and capacity during the test. The impedance spectra exhibit a smaller increase of span and only small changes in their shape during aging, see figure 7.

Table 1. Parameterization of EIS measurement.

Parameter	Value
Current amplitude AC	$0.28A$
Current amplitude DC	$0A$
Measured frequencies	49
Frequency range	$11mHz...2.85mHz$
Frequency distribution	$10freq/dec, f \geq 0.1Hz$
	$3freq/dec, f \leq 0.1Hz$
Test duration	approx. $20min$

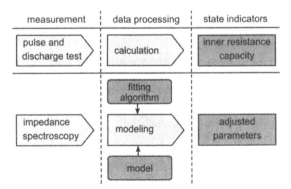

Figure 1. Process for determining state indicators from measured data for both pulse and discharge test and impedance spectroscopy.

3

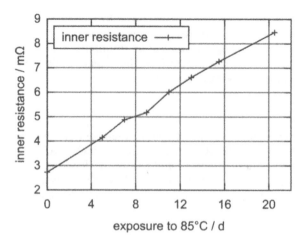

Figure 2. Development of the inner resistance (100% SOC, 5*s*) in the calendar aging experiment at 85°C.

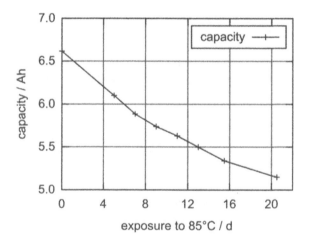

Figure 3. Development of the capacity in the calendar aging experiment at 85°C.

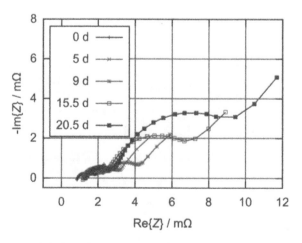

Figure 4. Development of the impedance spectra in the calendar aging experiment at 85°C.

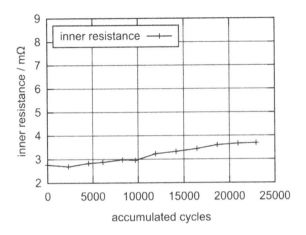

Figure 5. Development of the inner resistance (100% SOC, 5s) in the cycle aging experiment.

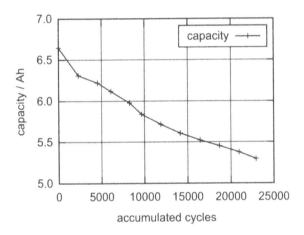

Figure 6. Development of the capacity in the cycle aging experiment.

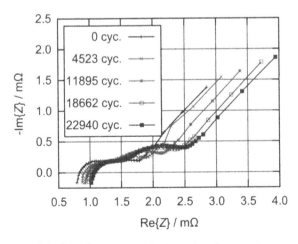

Figure 7. Development of the impedance spectra in the cycle aging experiment.

4 MODELING IMPEDANCE SPECTROSCOPY DATA

The challenge now is to gather the information hidden in the measured impedance curves (figures 4 and 7). This can be done using a well adjusted model in combination with an accurate fitting algorithm. Compared to the results of the pulse and discharge tests, which are directly available in a scalar form, this modeling process is necessary to derive scalar parameters from the impedance spectra.

4.1 *Equivalent circuit model*

To grasp the information hidden in an impedance spectrum and extract state indicators it is necessary to convert the impedance curves into a scalar description. As widely investigated in literature (Boukamp 2004), (Buller 2003), (J.B. Gerschler and Sauer 2006), (M.E. Orazem and Membrino 2002), (U. Trltzsch and Trnkler 2006) appropriate models for describing impedance spectra of li-ion cells are equivalent circuit models containing lumped and distributed elements. A simple equivalent circuit was introduced by (J.B. Gerschler and Sauer 2007) and contains one or more parallel connections of resistors and capacitors in series. This model is able to approximately capture the shape of the impedance spectra.

However, when having a detailed look on the semicircles of the spectra they are obviously of depressed shape. Due to mathematical reasons they cannot be fitted by lumped elements with high accuracy. Therefore, in the model of choice for this application the lumped elements are replaced by distributed elements. This model is depicted in figure 8. The mathematical equation (1) describes the terminal behavior of this equivalent circuit model. The impedance of this circuit is a non-linear equation with 10 parameters. These 10 parameters are the component values of the electrical circuit elements: L_1 represents the inductive parts such as current collectors and connections within the cell. R_2, R_3 and R_4 are ohmic resistors and describe the ohmic resistances of electrolyte, current collectors and electrodes. Q_3, Q_4, Q_5 and α_3, α_4, α_5 describe the CPE elements, that enable the model to adapt to depressed semicircles and act effectively like capacitors representing the electrode and diffusion behavior (Jossen 2009), (Zoltowski 1998).

$$Z(\omega) = L_1(j\omega) + R_2 + \sum_{n=3}^{4} \frac{R_n}{1 + R_n Q_n (j\omega)^{\alpha_n}} + \frac{1}{Q_5(j\omega)^{\alpha_5}} \tag{1}$$

4.2 *Particle swarm optimization fitting algorithm*

Fitting the 10 model parameters (L_1, R_2, R_3, R_4, Q_3, Q_4, Q_5, α_3, α_4, α_5) to the measured impedance spectrum requires an appropriate optimization algorithm. The goal of the parameter adjustment is to minimize the overall error, i.e. the weighted sum of the squared difference between measured and modeled points. Since the spectra change massively over age the range covered by the parameter values is large and hence it is difficult to find universally valid starting parameters. But this is a prerequisite when using a gradient descent algorithm (Boukamp 2004), (Nelder and Mead 1965) since otherwise it is very likely to run into local minima. This is by passed by utilizing the non-deterministic PSO algorithm (Eberhart and Y 2001), (Kennedy and Eberhart 1995) where the entire optimization space is covered in the initial step.

We will now introduce the nomenclature for the PSO algorithm: The swarm consists of an arbitrary number of particles. Each particle represents a 10 dimensional vector of the parameters to be fitted. These parameters are the component values of the electrical equivalent circuit model and therefore each particle represents a possible solution of the optimization problem, i.e. a fitted impedance spectrum. Each parameter is restricted to a one-dimensional interval. The choice of the lower and upper limits is mainly due to physical restrictions, e.g. all component values must be non-negative. The measured impedance

6

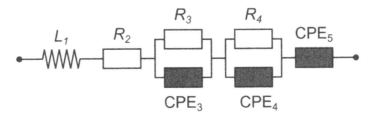

Figure 8.   Equivalent circuit RCPE model.

Data:
    N = number of particles
    D = dimension of particle
    I = number of iterations
    PSO parameters w, c_1 and c_2
    boundary conditions $\vec{x}_{min}$, $\vec{x}_{max}$ and $\vec{v}_{max}$
    error function $f : \mathbb{R}^D \to \mathbb{R}$, $\vec{x} \mapsto f(\vec{x})$

Result:
    best parameter set found so far $\vec{g}_{best}$

begin
    INITIALIZATION
    for each *particle n* **do**

$$\vec{x}_n = (\vec{x}_{max} - \vec{x}_{min}) \cdot \mathrm{rand}_{[0,1] \times D} + \vec{x}_{min}$$

$$\vec{v}_n = \vec{v}_{max} \cdot \mathrm{rand}_{[-1,+1] \times D}$$

$$\vec{p}_{best,n} = \vec{x}_n$$

    end

$$\vec{g}_{best} = \vec{p}_{best,k} \quad \text{with } f(\vec{x}_k) \le f(\vec{x}_l), \forall l \in N$$

    ITERATION
    for $i = 1 : I$ **do**
        for each *particle n* **do**
            if $f(\vec{x}_n) < f(\vec{p}_{best,n})$ **then**

$$\vec{p}_{best,n} = \vec{x}_n$$

                If $f(\vec{p}_{best,n}) < f(\vec{g}_{best})$ **then**

$$\vec{g}_{best} = \vec{p}_{best,n}$$

            end
        end

$$\vec{v}_{n,i+1} = w \cdot \vec{v}_{n,i}$$

$$+ \mathrm{rand}_{[0,1] \times D} \frac{c_1}{\Delta t} (\vec{p}_{best,n,i} - \vec{x}_{n,i})$$

$$+ \mathrm{rand}_{[0,1] \times D} \frac{c_2}{\Delta t} (\vec{g}_{best,i} - \vec{x}_{n,i})$$

        apply boundary conditions for all $\vec{v}_{n,i+1}$

$$\vec{x}_{n,i+1} = \vec{x}_{n,i} + \vec{v}_{n,i+1} \cdot \Delta t$$

        apply boundary conditions for all $\vec{x}_{n,i+1}$
        end
    end
end

Figure 9.   Schematic description of the PSO algorithm.

spectrum is compared to the fitted impedance curve represented by each particle by evaluating the error function, which leads to a scalar value of the overall error. This is a measure for the quality of the fit given by this particular particle. During the optimization process the particles move through the optimization space. The personal best position of each particle, which is the parameter set that fitted the measured values best, is stored in the memory of this particle. During the iterations the best known position that any particle has ever achieved is stored and if necessary updated in the global memory of the swarm. For deriving direction and strength of each particle's movement a velocity vector is calculated in each iteration step.

The pseudo code description of the PSO algorithm can be found in algorithm 1. In a first initialization step all particles are allocated randomly covering the whole optimization space. For each particle the personal best position is derived and the global best position is determined. Subsequently, the velocity vector is calculated for each particle and out of this the new positions are calculated. To ensure that all parameters stay within the optimization space, boundary conditions are monitored and if necessary a correction is performed. This loop is carried out until a termination criterion is reached. Typical termination criteria include having performed a maximum number of iterations or falling below a arbitrarily small global error.

The intelligence of the swarm is reflected in the velocity vector in the i-th iteration step that is used for the computation of each element's position in the next iteration step, see equation (2).

$$\vec{v}_{n,i+1} w \cdot \vec{v}_{n,i} + rand_{[0,1] \times D} \frac{c_1}{\Delta t} (\vec{p}_{best,n,i} - \vec{x}_{n,i}) + rand_{[0,1] \times D} \frac{c_c}{\Delta t} (\vec{g}_{best,n,i} - \vec{x}_{n,i}) \tag{2}$$

Where $\vec{v}_{n,i}$ is the velocity from the previous iteration step, $\vec{p}_{best,n,i}$ is the personal best position so far, $\vec{g}_i$ is the global best position so far and $\vec{x}_{n,i}$ is the current position of the n-th particle. The time for one iteration step is Δt which is in this case equal to one. The velocity vector consists of three additive terms. First, the previous velocity from the former iteration step is kept. Additionally, the particle is attracted to its personal best position known from former iteration steps. Finally, the particle is attracted to the global best position of the swarm. All three components are weighted by constant values w, c_1 and c_2 that are randomly disturbed to ensure variability. The influence and knowledge of the swarm's and each particle's history enables a higher flexibility in searching the whole optimization space and leads to a much better fitting of the measured impedance data, compared to common gradient descent algorithms.

Figure 10 illustrates the extremely good performance of the PSO algorithm for fitting impedance spectra.

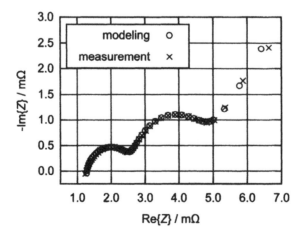

Figure 10.   Performance of PSO fitting.

5 RESULTS

The aging parameters of the pulse and discharge tests for both aging experiments were given in section 3. The capacity (figures 3 and 6) shows a continuous decrease over time. The increase of the inner resistance differs between both experiments (figures 2 and 5), with a stronger increase for the calendar aging.

For the evaluation of the impedance spectra the RCPE model (figure 8) was fitted to the measured data using the PSO algorithm. Algorithm and underlying model are capable of modeling the spectra with a very high accuracy as can be seen in figure 10. The impedance spectra (figures 4 and 7) were reduced to sets of 10 parameters.

Due to aging some of the parameters show a significant variation, others remain fairly constant. Since the current collectors do not degrade and the electrolyte resistance does not increase, L_1 and R_2 remain constant during the aging process, as well as the parameters Q_3, Q_4 and including the exponents α_3, α_4, α_5.

During calendar aging mainly the second depressed semicircle changes over time, which is represented by R_4. The other ohmic parameters do not show significant trends, see figure 11. The parameter Q_5 (figure 12), which corresponds to diffusion phenomena in the cell, decreases significantly with age.

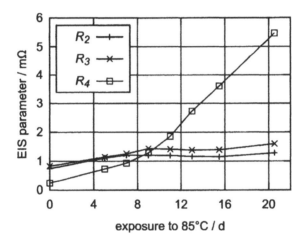

Figure 11. Development of the EIS parameters R_2, R_3, R_4 (100% SOC) in the calendar aging experiment at $85°C$.

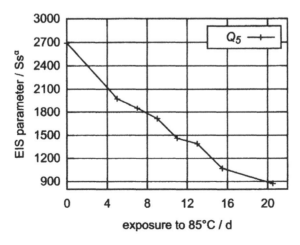

Figure 12. Development of the EIS parameter Q_5 (100% SOC) in the calendar aging experiment at $85°C$.

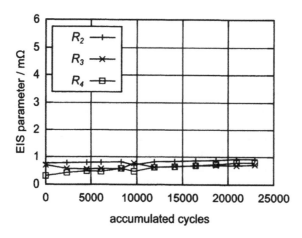

Figure 13.   Development of the EIS parameters R_2, R_3, R_4 (100% SOC) in the cycle aging experiment.

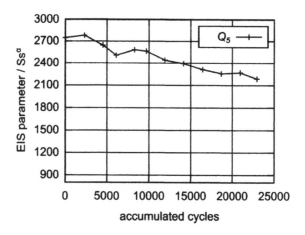

Figure 14.   Development of the EIS parameter Q_5 (100% SOC) in the cycle aging experiment.

During cyclic aging, resistors R_2, R_3, R_4 exhibit only a mild increase (figure 13), this corresponds fully to the behavior of the inner resistances in the pulse and discharge test. The parameter Q_5 decreases similar to the behavior in the calendar aging experiment (figure 14).

So we can summarize that the increase of inner resistance found in the pulse and discharge test corresponds to a strong increase of the EIS parameter R_4, while the decrease in the cell's capacity corresponds to a decrease in the EIS parameter Q_5. So the EIS parameters R_4 and Q_5 are identified as aging indicators and supply information about inner resistance and capacity of the li-ion cells.

6   CONCLUSION

The proposed equivalent circuit model combined with the Particle Swarm Optimization algorithm is capable of fitting the impedance spectra of the li-ion cells with high accuracy. The modeling stays constant over the lifetime of the cells while the measured curves change their span and shape significantly. We found significant aging indicators among the adjusted EIS parameters that are capable of describing the cells SOH. The results of the aging experiments show that EIS could be used in future to provide valuable information about inner resistance and capacity of li-ion cells requiring a significant lower measurement time.

ACKNOWLEDGEMENTS

We would like to thank Andreas Gruhle for helpful discussions concerning the technical and electrochemical part and for providing us the possibility of using his laboratory and equipment. We would also like to thank Christoph Fischer for insightful discussions about the PSO algorithm and for help with the programming.

REFERENCES

Boukamp, B. (2004). Impedance spectroscopy, strength and limitations. *Technisches Messen 71*, 454–459.

Broussely, M., Biensan, P., F. B, P. B, S. H, K. N and Staniewicz, R.J. (2005). Main aging mechanisms in li-ion batteries. *Journal of Power Sources 146*, 90–96.

Buller, S. (2003). Impedance-based simulation models for energy storage devices in advanced automotive power systems.

Eberhart, R. and S. Y (2001). Particle swarm optimization: Developments, applications and resources. *Proceedings of the Congress on Evolutionary Computation*, 81–86.

Gerschler, J.B., A. H and Sauer, D. (2006). Investigation of cycle-life of lithium-ion batteries by means of electrochemical impedance spectroscopy. *Technische Mitteilungen 99*, 214–220.

Gerschler, J.B., Kowal, J., M. S and Sauer, D.U. (2007). High-spatial impedance-based modeling of electrical and thermal behaviour of lithiumion batteries—a powerful design and analysis tool for battery packs in hybrid electric vehicles. *Electric Vehicle Symposium (EVS 23)*.

Jossen, A. (2009). *Encyclopedia of Electrochemical Power Sources, chapter Batteries*. Elsevier.

Jossen, A. and Weydanz, W. (2006). *Moderne Akkumulatoren richtig einsetzen*. Ubooks Verlag.

Kennedy, J. and Eberhart, R. (1995). Particle swarm optimization. *IEEE International Conference on Neuronal Networks 4*, 1942–1948.

Macdonald, J.R. (2005). *Impedance spectroscopy* (2. edition ed.). John Wiley & Sons.

Nelder, J.A. and Mead, R. (1965). A simplex method for function minimization. *Computer Journal 7*, 308–313.

Orazem, M.E., P. S and Membrino, M.A. (2002). Extension of the measurement model approach for deconvolution of underlying distributions for impedance measurements. *Electrochimica Acta 47*, 2007–2034.

Sarre, G., P. B and Broussely, M. (2004). Aging of lithium-ion batteries. *Journal of Power Sources 127*, 65–71.

Trltzsch, U., O. K and Trnkler, H.-R. (2006). Characterizing aging effects of lithium ion batteries by impedance spectroscopy. *Electrochimica Acta 51*, 1664–1672.

Zoltowski, P. (1998). On the electrical capacitance of interfaces exhibiting constant phase element behaviour. *Journal of Electroanalytical Chemistry 443*, 149–154.

Lecture Notes on Impedance Spectroscopy – Kanoun (ed)
© *2012 Taylor & Francis Group, London, ISBN 978-0-415-69838-2*

Effects of soil properties on steel samples as studied with electro-chemical impedance measurements

Rudolf Holze

Institute of Chemistry, Chemnitz University of Technology, Chemnitz, Germany

ABSTRACT: The influence of corrosion activity of various types of soil on steel employed in liquefied gas tanks as estimated with a standard method is measured both with linear electrode potential scans and electrode impedance measurements. Results agree closely indicating no effect of soil properties as classified, this is in agreement again with the very similar pH-values of aqueous soil extracts and their ionic conductivity.

Keywords: Impedance measurement, corrosion, gas tank

1 INTRODUCTION

Tanks made from steel buried underground and used for storage of liquefied gas are protected against corrosion by epoxy coatings. Regular inspection of the integrity is mandated by law in most countries, details pertaining to regulations in e.g. Germany are provided in various texts (e.g. DIN 4681-1; DIN 4681-3). Currently an inspection of tanks with <3 to capacity is obligatory every ten years provided the tank is not exposed to particularly destructive or otherwise challenging environmental conditions. This examination has been performed in the past predominantly by inspection of the inner surface of the steel tank. This procedure is cumbersome, it requires transfer of the liquefied gas into a separate container, generous ventilation of the emptied container and visual surface inspection by a person who has to climb into the tank. Obviously this procedure is a risky one. In addition it is rather dubious how corrosion—which will proceed on the outer surface in case the coating is defective—shall be detected. From the content of the tank hardly any corrosive attack can be expected, all typical environmental conditions (humidity, electrolyte, oxygen or protons) necessary for corrosion are completely absent. On the contrary the major risk of corrosion is present on the outer surface. Any damage of the coating may permit access of exactly these corrosion-enabling substances to the steel surface. Thus any method capable of detecting reliably damages of the epoxy coating are welcome. When they can be employed without removing the tank from its site they will be particularly helpful because unnecessary expenses and risks can be possibly completely avoided. TÜV SÜD Industrie Service GmbH, Munich, Germany, has developed a method based on a DC-measurement of the electric resistance between the steel of the tank and a probe (spear) placed in the ground close to the tank (Registered design/trademark 20 2005 009 752.7). A flow of electric current easily indicates damage to the integrity of the coating. As shown in [FIG: 1] in case of a perfect coating the resistance should be infinite, current should be zero. Actually a small residual current because of leakage at cable insulations etc. may be expected. There are two interfaces to be considered. The interface between the spear and the (wet) soil (1) exhibits possibly typical properties of an electrochemical double layer. The second interface (2) is a completely insulating one provided the epoxy coating is perfect. At first glance—because of thenon-Ohmic behaviour of interface (1)—application of an AC-voltage as widely done in measurements of electrolytic conductivity seems to be the suitable approach. Because of the very large area of interface (2) (depending on size and shape of the tank several square meters) a considerable capacitance

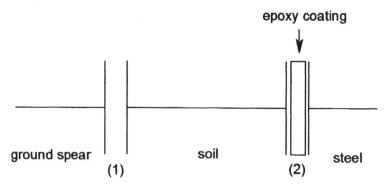

Figure 1.   Simplified equivalent circuit of the arrangement for measuring the integrity of a tank coating.

is established making detection of any imperfections in the coating resulting in the formation of an electro-chemical double layer and flow of large capacitive electric current impossible. With a DC-voltage the non-linear behavior of the interface (1) has to be taken into account. A voltage of $75V$ as currently employed in the suggested instrument will be enough to cause a measurable current in case of a defective coating, the potential drop at interface (1) will be overcome easily The same applies to interface (2) in case of a defect. In the latter case a modified equivalent circuit is more appropriate with DC-current flowing across interface (3) at the defect site. In a simplified consideration of the potential drop between the spear and the steel of the tank a drop of about one to two volts appears across interface (1), in case of average soil conditions and the rather low flow of current the potential drop in the soil will be small. In a simple experimental setup using the spear and a piece of coated wire with only a small metal surface area exposed to the soil a resistance of average soil of about $10k\Omega$ could be estimated, the resistive equivalent of the spear/soil interface was negligibly small (unpublished results). The resistance assigned to the interface soil/defect ((3) in FIG: 2) obviously depends on the contact area. In a typical experiment a resistance of $43k\Omega$ was measured with a contact area of $A = 0{,}096mm^2$. Taking into account safety margins a minimum value of $120k\Omega$ has been suggested as being sufficiently indicative of an integer coating. Although the corresponding size of a defect in the coating will be extremely small it was considered to be necessary to estimate the rate of corrosion and the time necessary to yield a hole in the wall of the tank. Thus measurements of rates of corrosion were performed.

Because the tanks are embedded in soil of somewhat variable properties these measurements should be performed with soil samples of known differences in corrosion aggressiveness. Currently in Germany this property is expressed as a characteristic determined by the "Wenner"-procedure obtained by measurements of soil conductivity once again based on application of DC-voltages between two spears in the ground. The specific soil resistance ς is calculated according to

$$\varsigma = 2 \cdot \pi \cdot d \cdot R \tag{1}$$

with d being the distance in cm and R being the resistance in Ω. The measurement is performed at different distances ranging from 20 to $80cm$. This way current distribution is taken into account approximately. Specific soil resistance values and the assigned characteristics are collected in Table 1. In case of substantial variations of ς with distance (and the equivalent depth) inhomogeneities have to be assumed, a further characteristic number is determined as collected in Table 2. The characteristics are used to determine the need for cathodic corrosion protection, a larger soil resistance being indicative of a lower aggressiveness. Although these considerations are apparently purely empirical it seems to be sure to assume, that higher conductivity is taken as a strong indicator of higher concentration of ions. This in turn is associated with higher aggressiveness. Differences in properties of ions are not taken into

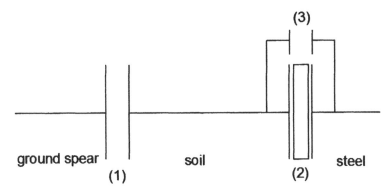

Figure 2. Simplified equivalent circuit of the setup in case of a defective tank coating.

Table 1. Characteristics according to Wenner-procedure.

Specific soil resistance values (Ωcm)	Characteristic
<10000	18
10000 .. 13000	12
13000 .. 20000	6
>20000	2

Table 2. Homogeneity of soil.

Resistance ratio: $\varsigma_{max}/\varsigma_{min}$	Characteristic
<1.5	0
1.5 .. 2	3
2 .. 3	6
3 .. 4	9
>4	12

Table 3. Soil resistance values of investigated samples.

	Location in coordinates		Soil resistance values/distance					
			80		40		20	
	N	E	(Ω)	(Ωcm)	(Ω)	(Ωcm)	(Ω)	(Ωcm)
sample 0	12.92802	50.83832	n.a	n.a	n.a	n.a	n.a	n.a
sample 1	11.99774	48.85821	14.88	7480	24.9	6258	59.4	7464
sample 2	12.19052	49.59023	4.94	2483	11.32	2845	30	3770
sample 3	12.12112	48.49847	28.9	14527	66	16588	187	23499
sample 4	12.35802	50.56183	74.4	37398	208	52276	270	33929

account. The fact, that changes of pH may cause considerable changes of conductance even at almost constant total ion concentration is also overlooked. The sum of both characteristics (from Tables 1 and 2) is calculated, at a value >18 cathodic corrosion protection is recommended. Soil samples were collected at five different sites in Germany, pertinent data including location are collected in Table 3. The studied soil samples showed a broad variation of

Table 4. Soil extract properties.

Sample	pH/–	Initial conductance $(mS \cdot cm^{-1})$	Conductance after filtration $(mS \cdot cm^{-1})$	Conductance after addition of K_2SO_4 $(mS \cdot cm^{-1})$
0	6,4	0,3	0,63	6,4
1	6,6	0,22	0,63	6,44
2	5,6	0,16	0,26	6,41
3	6,4	0,21	0,38	6,52
4	6	0,15	0,17	6,42

ς implying substantial differences in expected aggressiveness. Because the tank examination procedure discussed above must yield predictive data for all soil conditions these samples were employed in comparative corrosion measurements.

2 EXPERIMENTAL

Impedance measurements and linear electrode potential scans at a scan rate of $5mV/s$ were performed in a three-electrode H-cell. The steel sample (Steel STE 355) was machined into a disc of $1.5cm$ diameter and attached to a lucite holder fitted to the central port of the cell. The sample circumference and the steel-lucite joint were carefully covered with PTFE tape, only the frontal surface of $1.7cm^2$ was exposed to the electrolyte solutions. It was polished on abrasive paper 1000 grade and with a slurry of $\alpha-Al_2O_3$ ($13\mu m$). A platinum wire and a saturated calomel electrode in separate compartments separated from the main compartment with low-porosity glass frits were employed. For potential scans and electrochemical impedance measurements (EIM) a potentiostat Solartron SI 1287 connected to a frequency response analyzer SI 1255 interfaced to a PC with custom-developed software was used. Impedance measurements were carried out at the spontaneously established open circuit (corrosion) potential (OCP) with a modulation amplitude of $5mV$ in a frequency range from $0.1Hz$ to $100kHz$. Evaluation of the impedance data was performed with Boukamp software version 2.4. All experiments were performed at room temperature with electrolyte solutions saturated with air. Electrolyte solutions were prepared from aqueous soil extracts obtained by drying $50g$ of soil for $24h$ at $T = 100°C$ (weight loss about ca. 10–20%wt.) and subsequent dispersion in $100ml$ distilled water. After $24h$ exposure conductance and pH-value were measured, results are collected in Table 4. Addition of K_2SO_4 (p.A., Ferak) resulting in a concentration of $30mM$ yielded conductance data collected in Table 1 sufficient for impedance measurements. In a second set of experiments (for details see below) $KClO_4$ (p.A., Merck) was added up to the same molar concentration.

3 RESULTS AND DISCUSSION

Electrolyte solutions prepared by extraction of soil samples showed conductance values measured initially somewhat lower than those observed after filtration indicating slow establishment of any equilibration processes resulting in the release of ions. After about $24h$ a stable value was observed. These values were too low for reliable and reproducible impedance measurements, thus the described addition of K_2SO_4 was performed based on the assumption that sulphate anions do not specifically support corrosion. As a cross check addition of $KClO_4$ resulting in the same added electrolyte concentration was done based on the frequently observed almost complete absence of specific adsorption of this anion on many metals (Holze and Fischer 1989), (Kania and Holze 1998). Observed corrosion potentials E_{corr} did not show any correlation with "Wenner"-characteristics, they carry no information about corrosion rates anyway. Linear potential scans yielded Tafel-plots, a representative plot in FIG: 3 shows the salient data E_{corr} and i_{corr} as obtained from these plots. Impedance measurements performed with exactly the same sample, solution etc. resulted in data based on the equivalent circuit shown in FIG: 4. A typical Nyquist diagram illustrates the average quality of the obtained fit. Representative results of both methods are collected in Table 5. Taking into account the

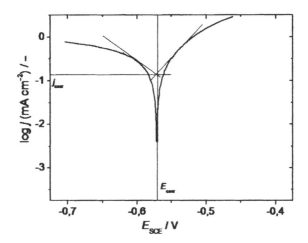

Figure 3.   Tafel plots for a steel electrode in electrolyte solution, for details see text.

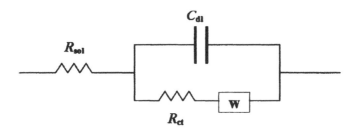

Figure 4.   Equivalent circuit used in evaluation of impedance measurements.

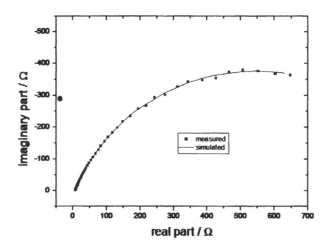

Figure 5.   Nyquist diagrams of a steel electrode, symbols: measured data, solid line simulation based on fit.

Table 5. Corrosion data with electrolyte solution containing added K_2SO_4.

Soil sample 0	Soil sample 1	Soil sample 2	Soil sample 3	Soil sample 4
Linear electrode potential scan results				
$E_{corr} = -505mV$	$E_{corr} = -550mV$	$E_{corr} = -560mV$	$E_{corr} = -590mV$	$E_{corr} = -606$
$|i_{corr}| = 28\mu A$	$|i_{corr}| = 25\mu A$	$|i_{corr}| = 25\mu A$	$|i_{corr}| = 18\mu A$	$|i_{corr}| = 33\mu A$
Electro chemical impedance measurement results				
$R_{ct} = 697\Omega$	$R_{ct} = 183\Omega$	$R_{ct} = 480\Omega$	$R_{ct} = 557\Omega$	$R_{ct} = 142\Omega$
$|i_0| = 18\mu A$	$|i_0| = 68\mu A$	$|i_0| = 26\mu A$	$|i_0| = 22\mu A$	$|i_0| = 88\mu A$
$CR_{min} = 0,14mm/a$	$CR_{min} = 0,14mm/a$	$CR_{min} = 0,2mm/a$	$CR_{min} = 0,14mm/a$	$CR_{min} = 0,16mm/a$
$CR_{max} = 0,75mm/a$	$CR_{max} = 0,6mm/a$	$CR_{max} = 0,2mm/a$	$CR_{max} = 0,46mm/a$	$CR_{max} = 0,86mm/a$

Table 6. Corrosion data with electrolyte solution containing added $KClO_4$.

Soil sample 0	Soil sample 1	Soil sample 2	Soil sample 3	Soil sample 4
Linear electrode potential scan results				
$|i_{corr}| = 5\mu A$	$|i_{corr}| = 3\mu A$	$|i_{corr}| = 4\mu A$	$|i_{corr}| = 3\mu A$	$|i_{corr}| = 2\mu A$
Electrochemical impedance measurement results				
$R_{ct} = 1620\Omega$	$R_{ct} = 2765\Omega$	$R_{ct} = 1026\Omega$	$R_{ct} = 2074\Omega$	$R_{ct} = 1600\Omega$
$|i_0| = 8\mu A$	$|i_0| = 5\mu A$	$|i_0| = 12\mu A$	$|i_0| = 6\mu A$	$|i_0| = 8\mu A$
$CR_{min} = 0,04mm/a$	$CR_{min} = 0,02mm/a$	$CR_{min} = 0,02mm/a$	$CR_{min} = 0,02mm/a$	$CR_{min} = 0,01mm/a$
$CR_{max} = 0,08mm/a$	$CR_{max} = 0,13mm/a$	$CR_{max} = 0,09mm/a$	$CR_{max} = 0,08mm/a$	$CR_{max} = 0,06mm/a$

substantial scattering of data—which comes as no surprise because of the very realistic composition of the electrolyte solution containing even unknown amounts of biological matter with unknown activity—minimum and maximum rates of corrosion CR were calculated as displayed in Table 5. No influence of the soil composition can be observed. Apparently the discussed characteristics based on the "Wenner"-procedure fail to show predictive or indicative value. Taking into account the basic flaws of the procedure as already discussed above this is hardly a surprise. The variation of pH-value between the investigated samples is small, although significant, but again it shows no correlation with the corrosion behavior nor the "Wenner"-characteristic. To examine a conceivable influence of the electrolyte added to enhance ionic conductivity of the soil extract $KClO_4$ was used instead of K_2SO_4 in a second set of measurements. Calculated rates of corrosion are substantially lower, in a few cases by an order of magnitude. Apparently sulphate ions are not as innocent at all contrary to experience with its behavior in other electrochemical systems. The latter values of corrosion rates are similar to those rates observed with tanks where inspections showed corrosion at defective coating sites. Taking into account the thickness of the wall of the tank within the inspection interval of ten years no leaks because of corrosive wall perforation has to be expected.

ACKNOWLEDGMENTS

Stimulating discussion and generous support by TÜV Süd (Jens Helmstedt and Winfried Schock) and assistance by performing and evaluating electrochemical measurements by Stefan Herkert is gratefully appreciated.

REFERENCES

Holze, R. and Fischer, G. (1989). DECHEMA-Monographie. Number 117.
Kania and Holze, R. (1998). GDCh-Monographie. Number 14. Frankfurt: GDCh.

Lecture Notes on Impedance Spectroscopy – Kanoun (ed)
© *2012 Taylor & Francis Group, London, ISBN 978-0-415-69838-2*

Magnetic field effect on hydrogen evolution activity of electrodeposited Ni-Mo coatings

Omar Aaboubi
Chemistry Department, Reims Champagne Ardenne University, Reims, France

ABSTRACT: Enhancement of the catalytic activity of NiMo alloys for hydrogen evolution reaction (HER) has been observed when the alloy is electrodeposited under magnetic field B, control. Using cyclic voltammetry at low scan rate and electrochemical impedance spectroscopy (EIS) measurements, the significant modification of the catalytic activity of the alloy has been observed in acidic water, in basic solution at $25°C$ and in industrial conditions (hot alkaline solution). For the NiMo alloy electrodeposited under applied magnetic field, the stationary current-potential curves show high current densities, and more pronounced potential shift to the positive domain. The EIS measurements corroborate the stationary observations and show a decrease of the polarisation resistance, the charge transfer resistance and an increase of the double layer capacitance according to the modifications of the deposit properties improved by the forced magnetohydrodynamic convection.

Keywords: Ni-Mo alloy, magnetic field, electrodeposition, electrochemical impedance, hydrogen evolution reaction, water electrolysis

1 INTRODUCTION

Hydrogen as an energy carrier can play an important role as an alternative to conventional fuels, provided, its technical problems of production, storage and transportation, can be resolved satisfactorily and the system cost could be brought down to acceptable limits (Dunn 2002), (Dincer 2002), (A. Hugo 2005), (H.J. Russell 2007).

One of the practical ways to obtain hydrogen is water electrolysis which will be the most efficient production process of hydrogen without any environmental problems (R. Khotari 2005).

The mean operating cost of hydrogen evolution reaction (HER) is the cost of electricity. Hence, research and development efforts have been recently focused on the minimizing ohmic drop, lowering the overpotential through the improvement of the cell and electrode design and using electrode with high catalytic activity. Electrochemical alloy deposition is widely employed to produce new materials for specific applications (mechanics, physics or chemical applications) such as Nickel alloys due to their catalytic properties and for which, the addition of molybdenum improvesits catalytic efficiency during the HER (J.M. Jaksic 2000), (L. Birry 2004), (S. Martinez 2006), (N.V. Krstajic 2008), (Q. Han 2010), (M.A. Dominguez-Crespo 2005), (E. Navarro-Flores 2005).

As is well known, application of an external magnetic field during electrochemical reaction generates various modifications, essentially due to the induced magnetohydrodynamic (MHD) forced convection (R. Aogaki 1975), (Fahidy 1983), (O. Aaboubi 1990), (J. Lee 1995), (N. Leventis 1998), (G. Hinds 2001). This effect depends on the magnetic field direction in regards to the electrode surface orientation. For a homogeneous external magnetic field B parallel to the working electrode surface (with diameter d), the current density is perpendicular to the B direction and a Lorentz force is generated. The field induced motion of the electrolyte near the electrode and consequentially, when the natural convection can be neglected, the diffusion limiting current is $I_l \propto d^{5/3} C^{4/3} B^{1/3}$ (O. Aaboubi 1990), (N. Leventis 1998) where C is the bulk concentration of the electroactive species.

If the magnetic field is perpendicular to the electrode surface and the electroactive species under mass transport control has paramagnetic properties, a stirring of the liquid near the electrode is induced and consequentially, the electrolysis current increased. For the diffusion limiting current the relationship $I_l \propto d^{7/4} C^{5/4} B^{2/3}$ was obtained for a vertical electrode (K.L. Rabah 2004).

In our laboratory, we have shown previously that during the metal or alloys electrodeposition, the application of B improves the deposit quality (grain size, surface homogeneity and roughness) (O. Devos 1998), (K. Msellak 2004). Devos et al. reported that some changes of the surface morphology and the preferred orientation of the nickel grain under B control (O. Devos 1998). The authors attributed their modifications to the MHD convection generated near the electrode which increases the inhibiting species diffusion flux. Fahidy wrote a no table review in the field (Fahidy 2001) and reported that, the surface roughness and the three-dimensional deposit structure may be affected by the MHD convection (K. Msellak 2004). Msellak et al. (K. Msellak 2004) for NiFe alloy and Zabinski et al. (P.R. Zabinski 2009) for Co-P alloy, showed the decrease of the grain size of the deposits under an applied magnetic field. Zabinski et al. consider that due to the MHD convection the morphology of the deposit was affected allowing to high activity of the Co-P alloy for HER (P.R. Zabinski 2009).

In the present work, the effect of applied homogeneous magnetic field parallel to the electrode was performed to enhance the catalytic activity of Ni-Mo electrodeposited coating for HER. The activity of the alloy was tested in acidic water, basic water at $25^\circ C$ and in industrial conditions (hot and concentrated NaOH solution). To avoid weighing down the text, we have chosen to discuss here only the results obtained in acidic water.

2   EXPERIMENTAL SECTION

For the Ni-Mo alloy electrodeposition, the chemical composition of bath and the experimental conditions are the same as that used in our previous work (O. Aaboubi 2010).

The thin Ni-Mo films were electrodeposited onto Pt working electrode (WE) by applying a constant potential value of $E = 1.20 V/CSE$ or a constant cathodic current density value of $J = 35 mA \cdot cm^{-2}$ during $30 min$. The chemical composition of prepared alloy was characterized by scanning electron microscopy in combination with energy dispersive X-ray microanalysis (SEM/EDX).

For the electrodeposition of the Ni-Mo deposits under applied magnetic field, the whole of the cell was inserted between the poles pieces of an electromagnet (SIGMA-PHI) coupled to a regular power supply (BOUHNIK). This magnet can generate a homogeneous magnetic field up to $1.2T$ for $7cm$ gap and $1.65T$ for $5cm$ gap. The working electrode was maintained in the vertical position facing parallel to the magnetic field direction. In this configuration, the field is perpendicular to the cuurent lines.

The HER activity of the Ni-Mo deposits was examined by means of the measurements of the stationary polarization curves coupled with EIS measurements using electrolytic solution of $0.5 mol \cdot L^{-1}$ of H_2SO_4 and controlled temperature of $25^\circ \pm 1C$. The current potential curve measurements were performed using a Potentiostat/Galvanostat (Voltalab-PGZ 100) monitoring by Voltamaster 4 software. A potential scan rate of $v = 1 mV \cdot s^{-1}$ was chosen to get stationary state values of the electrolysis current. The EIS measurements were carried out using a home-made potentiostat and a Frequency Response Analyser (Solartron 1250) controlled by Zplot 2.4 (Scribner Associates). To get an EIS measurement in the linear mode, $10 mVrms$ was used for sine wave potential amplitude for frequency range from $10 kHz$ to $1 mHz$.

3 RESULTS AND DISCUSSION

3.1 *Electrodeposition of Ni-Mo deposits*

The first step, Ni-Mo films were electrodeposited for different conditions of applied magnetic field. Figure 1 shows the experimental potential-time curves for the Ni-Mo deposits obtained with magnetic $B = 1T$ and without B and a constant cathodic current density value of $J = -35mA \cdot cm^{-2}$. For the two deposition conditions, a constant value of the potential was reach after rapid time amount. With magnetic field, the stationary potential value is less then that without B indicating the decrease of. As already shown (R. Aogaki 1975), (Fahidy 1983), (O. Aaboubi 1990), (J. Lee 1995), (N. Leventis 1998), (G. Hinds 2001), the MHD convection generated by the combination of the magnetic field perpendicular to the current lines near the electrode, increases the concentration of the electroactive species in the vicinity of the electrode allowing to the decrease of the overpotential of the alloy electrodeposition.

The composition of metallic elements in the deposit thus prepared was performed using the energy dispersive X-ray microanalysis (EDX). The means composition of 73% Ni and 27% Mo was obtained at $B = 0$ and of 75% Ni and 25% Mo was obtained at $B = 1T$. From this we can not consider some modifications of the alloy composition by applying external magnetic field.

3.2 *Polarization curves I(E)*

The HER activity of obtained alloys was evaluated using cyclic voltammetry (CV) measurements at low potential scan rate of $1mV \cdot s^{-1}$ and compared to that obtained from the stationary polarization points of EIS measurements. Figure 2 shows the comparison of the polarization curves $I(E)$ measured for the Ni-Mo alloy obtained under applied magnetic field of $B = 1T$ (so called: NiMo-B1T) and the alloy deposited without B (so called: NiMo-B0T) in aerated acidic water $(0.5MH_2SO_4)$. In both cases, the similar shape of all curves can be observed for the two deposits with significant increase of the current density (CD) values and a decrease of over potential were observed when the alloy was deposited under B. From Figure 2 we can also notice that the CV measurements obtained at scan rate of $1mV \cdot s^{-1}$ and the stationary values of the polarization points of the EIS measurements are practically

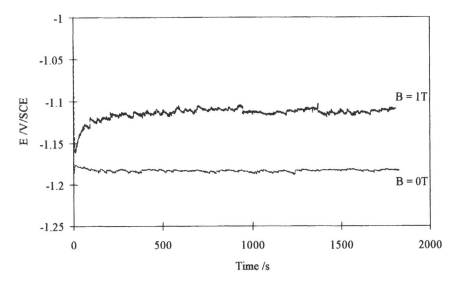

Figure 1.   Potential-time curves for electrodeposition of Ni-Mo alloy measured at applied cathodic current density of $I = 35mA \cdot cm^{-2}$, for $B = 0$ and $B = 1T$. Same solution conditions as described in [24].

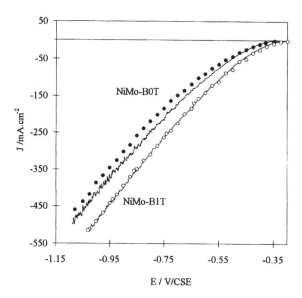

Figure 2. Typical cathodic polarization curves obtained during HER on NiMo-B0T deposit (full symbols) and on NiMo-B1T deposit (empty symbols). The lines correspond to CV measurements at $1mV \cdot s^{-1}$ and the symbols correspond to the stationary polarization points. $[H_2SO_4] = 0.5M$; temperature $T = 25^{\circ}C$ and vertical electrode (surface $A = 0.2cm^2$).

the same. We can also notice the decrease of the over potential of the HER on the deposit obtained under magnetic field. It is noteworthy that at a CD value of $I = 200mA \cdot cm^{-2}$, the overpotential value (see Eq. 1) is $\eta_{200} = 270mV$ using NiMo-B0T for HER, however the value is only $\eta_{200} = 217mV$ using NiMo-B1T. All of them confirm the magnetic field effect on the activity of the deposit.

3.3 *Tafel representation of the polarization curves*

The activity enhancement of the Ni-Mo alloy may be examined through the Tafel representation of the polarizations curves. In Figure 3, we have represented a set of the stationary Tafel curves obtained for the two deposits. The curves has been represented in term of the overpotential which is calculated at each polarization point from

$$\eta = E - E_0 - R_s \cdot I \qquad (1)$$

where E is the applied potential value, E_0 is the reset potential value, R_s is the solution resistance which is determined from the high frequency extrapolation of the electrochemical impedance data and the $R_s \cdot I$ term is the ohmic drop.

As previously reported (S. Martinez 2006), (N.V. Krstajic 2008), (Q. Han 2010), (M.A. Dominguez-Crespo 2005), (E. Navarro-Flores 2005) for HER study on various Ni-Mo alloys, two well-defined Tafel regions of the plots has been obtained for all solutions investigated here. In Figure 3, the obtained results show clearly that, the HER on the two samples exhibits a classical behavior of kinetically controlled electrochemical system and can be described by the Tafel relationship (Eq. 2)

$$\eta = a + b\log(I) \qquad (2)$$

where b is the Tafel slope and $a = b \cdot log(I_0)$ where I_0 is the exchange current density (ECD).

On Table 1, we have reported the obtained values of Tafel slopes and the ECD values of HER in acidic solution and for the NiMo-B0T and NiMo-B1T deposits. From Table 1, it is observable that the two kinds of deposit present nearly equivalent Tafel slopes in the low and

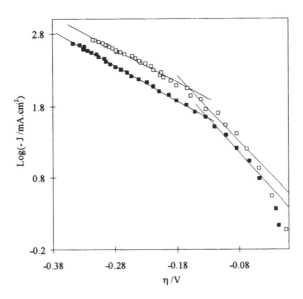

Figure 3. Tafel plots of the experimental stationary polarization curves for HER on NiMo-B0T deposit (full symbols) and on NiMo-B1T deposit (empty symbols) in acidic solution. $[H_2SO_4] = 0.5M$; temperature $T = 25°C$ and vertical electrode (surface $A = 0.2cm^2$).

Table 1. Typical electrochemical parameters deduced from linear Tafel plots of the stationary polarization curves obtained for HER in acidic solution.

Deposit	$b1/mV$	$I_{01}/mA \cdot m^2$	$b2/mV$	$I_{02}/mA \cdot m^2$
NiMo-B0T	103	2.40	208	10.3
NiMo-B1T	110	3.84	223	21.3

in the high overpotential domains. The observed difference between the slopes for the two deposits is not so significant to attribute them to any magnetic field effects. They may be due to the random determination errors or/and to the ohmic drop correction. As was noticed by Conway et al. (S. Martinez 2006), (J.O.M. Bockris 1993) for HER, very often two different Tafel slopes are observed: low slopes associated to charge transfer coefficient more then 0.5 at low CD values and the second slope which is associated to the high activation energy, appears at high CD values. Whereas for the ECD values the presence of applied B during alloy electrodeposition produces important modifications, i.e. the enhancement of the ECD values may reach more then 50% of the values. According to Bockris et al. (G.J. Brug 1984), the larger the ECD, the smaller will be the heat of activation of the reaction determined step (rds) and consequently, the reaction rate is higher. Hence, the observed increase of the ECD valuesmay be attributed to surface modifications and deposit quality changes induced by the MHD convection which takes place near the electrode during the alloy electrodeposition process. It shows clearly the increase of the activity for HER of the NiMo-B1T deposit. Hence, the presence of applied B during the electrodeposition process operates like some changes of the activation energy of the rds during the HER and consequently the hydrogen production is accelerated on the NiMo-B1T catalyst.

3.4 EIS measurements

To confirm the voltammetric observations with or without applied magnetic field, EIS measurements were performed at various potential values for the two deposits in acidic water.

Typical experimental EIS diagrams obtained on NiMo-B0T and NiMo-B1T deposits are represented in the Nyquist plane in Figure 4. For all potential values investigated here, the impedance diagrams present different characteristic parts, some of them changes dramatically when the alloys are deposited under applied magnetic field. As previously reported in (L. Birry 2004), (S. Martinez 2006) and (E. Navarro-Flores 2005), the EIS diagrams are composed of typical porous electrode response at high frequency domain, followed by two collapsed semi-circles in the low frequency domain. The high frequency part of the diagram being almost the same at all applied potential and also, seems to be not modified by applying magnetic field during the deposition process. It indicates that this part of the diagram is related to porosity of the deposit (L. Birry 2004). However, the evolution of the first high frequency capacitive loop, which corresponds to the charges transfer processes combined with the double layer charging process, is in accordance with the modifications of the stationary current density values observed before. The low frequency loop which is due to the adsorption process seems to by also modified by the magnetic field conditions of the deposits.

3.4.1 *Ohmic drop*
One of the recently development efforts have been focused on the minimizing the ohmic drop of electrolysis cell. This parameter is directly connected to the solution resistance R_s values. It is noteworthy that, the obtained values of R_s reported in Figure 6 show clearly the decrease of the solution resistance for the deposits obtained under magnetic field. This effect is directly related to the modification of the morphologic property of the deposit by the induced MHD convection. The consequence of this effect is the diminution of the ohmic drop during HER on NiMo-B1T deposit in comparison of the ohmic drop on NiMo-B0T deposit.

3.4.2 *Polarisation resistance*
When the frequency tends to zero value, the electrochemical impedance of the cell, Z tends towards a real number, namely the polarization resistance R_p, which corresponds to the inverse of the slope of the polarization curve, $I(E)$ at each stationary polarization point as follows

$$\omega \to 0: \quad Z = R_p = \frac{dE}{dI} \tag{3}$$

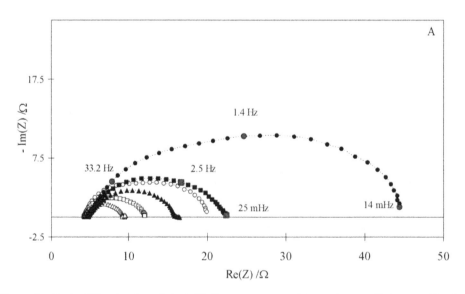

Figure 4. Typical Nyquist plot of the EIS diagrams measured for various applied potential values E, during HER on NiMo-B0T deposit (full symbols) and on NiMo-B1T deposit (empty symbols). $[H_2SO_4] = 0.5M$; Temperature $T = 25^\circ C$ and vertical electrode (surface $A = 0.2cm^2$). ($\bullet$ and $\circ$): $E = -0.35V/SCE$; ($\square$ and $\blacksquare$): $E = -0.40V/SCE$ and ($\blacktriangle$ and $\triangle$): $E = -0.45V/SCE$.

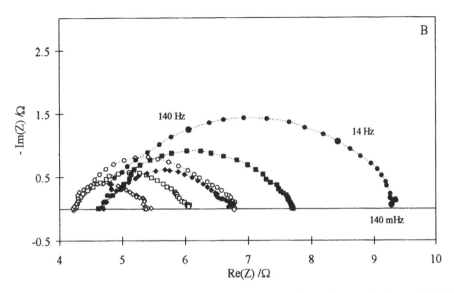

Figure 5. Typical Nyquist plot of the EIS diagrams measured for various applied potential values E, during HER on NiMo-B0T deposit (full symbols) and on NiMo-B1T deposit (empty symbols). $[H_2SO_4] = 0.5M$; Temperature $T = 25°C$ and vertical electrode (surface $A = 0.2cm^2$). (● and ○): $E = -0.60V/SCE$; (□ and ■): $E = -0.70V/SCE$ and (▲ and △): $E = -0.80V/SCE$.

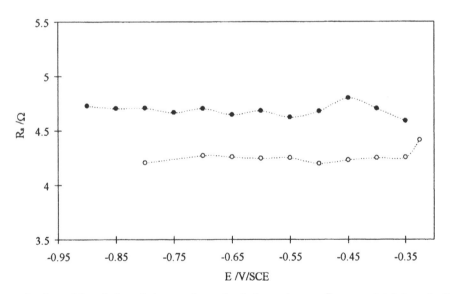

Figure 6. Potential evolution of the experimental solution resistance, R_s extrapolated from the EIS diagrams. NiMo-B0T deposit (full symbols) and NiMo-B1T deposit (empty symbols). Same conditions as Figure 4.

where $\omega = 2\pi f$ is the pulsation and f is the frequency. Figure 7 shows the comparison of the potential evolution of the experimental polarization resistance, R_p extrapolated from the impedance diagrams and the calculated value of the derivative, $\frac{dE}{dI}$ (Eq. 3) obtained from the stationary polarization curves at the same polarization point. It is noteworthy that practically, the same values are obtained for the two quantities. The comparison of R_p for the two

25

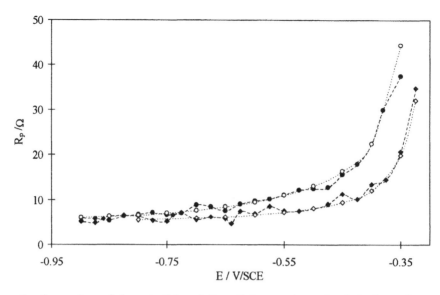

Figure 7. Comparison of the potential evolution of the experimental polarization resistance, R_p extrapolated from the EIS diagrams (empty symbols) and (dE/dI) calculated from the stationary polarization curves (full symbols). (● and ○): NiMo-B0T deposit and (♦ and ◊): NiMo-B1T deposit. Same conditions as Figure 4.

deposits reveals a decrease of the amplitude of R_p when the alloy is deposited under magnetic field control. The diminution is more remarkable at low polarization values however, for high potential values R_p seems to be constant for the two samples and the difference between them seems to be vanishing.

3.5 *Fitting procedure of the EIS diagrams*

To examine the magnetic field effect on EIS parameters, the experimental diagrams were analysed using a fitting procedure for different polarisation potential values. This method is based on the comparison of the experimental data with an equivalent circuit (EC) consisting of the solution resistance R_s, in series with combined parallel and series of R - constant phase elements, $CPE - R$ (Figure 8). The fitting procedure has been performed using commercial software ZSimpWin 3.21 (EChem Software). As an example, on Figure 9 we have represented in Nyquist plot, the comparison of the fitted diagrams and the experimental data obtained at $E = -0.40V/SCE$ for the two deposits. One can note the good agreement between the experimental and the fitted data in the frequency range between $10kHz$ and $25mHz$. We have also reported in Table 2 the detailed numerical results of the fittings, for the same polarisation potential values. Despite of the great number of adjustable parameters, many of them have been obtained with a reasonable statistical determination (i.e., the standard error is less then 8% for all fitted parameters).

3.5.1 *Charge transfer resistance*

In the high frequency domain the first loop of the diagram arises from coupling the electrons exchange reactions with the double layer charge phenomena. Usually the first process is represented by the charge transfer resistance, R_{ct} and the second by the double layer capacitance C_d. From this, the potential dependence of the charge transfer resistance, R_{ct} may be regarded in term of the potential evolution of the electrode kinetics. In Figure 10 we have represented the semi-logarithmic plot of R_{ct} as function of η, the data clearly show that alloy electrode position under B control causes a substantial lowering of R_{ct} in accordance with the increase

26

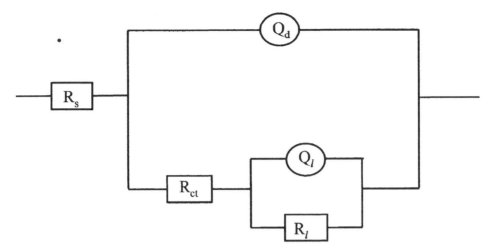

Figure 8. Equivalent circuit used for the fitting procedure of the experimental EIS data measured in acidic solution. R_s is the solution resistance, R_{ct} is the charge transfer resistance, Q_d is the high frequency CPE element, R_l is the low frequency resistance and Q_l is the low frequency CPE element.

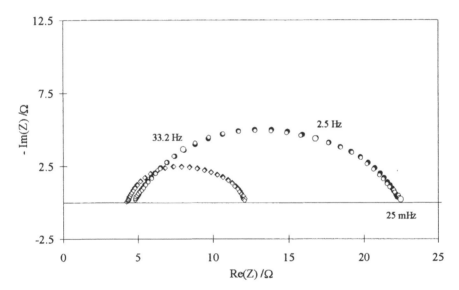

Figure 9. Comparison of the fitted (full symbol) and the experimental (empty symbol) impedance diagrams measured at $E = -0.40V/SCE$. (● and ○): NiMo-B0T deposit and (◆ and ◇): NiMo-B1T deposit. Same conditions as Figure 4.

of the stationary CD mentioned above. In accordance with the Tafel type of the electrode kinetics, straight lines have been obtained for the two deposits and exhibit the same slope value. We confirm here, the non modification of the Tafel slope obtained before from the stationary polarization curves (Table 1 and Figure 3).

3.5.2 *Double layer capacitance*
The measurements of the high frequency loop of the impedance diagrams on solid electrodes correspond rarely to the response of charge transfer resistance in parallel with a purely

Table 2. Typical electrochemical parameters obtained from fitting procedure of the EIS diagrams measured at E = 0.40 V/SCE for HER in acidic solution. The term St.er./%, corresponds to the standard error, in %, for each fitted parameters.

Deposits	Ni-MoB0T	Ni-MoB1T
R_s/Ω	4.70	4.25
St.er./%	0.59	0.38
$Q_d/\Omega^{1-\beta} \cdot F^{\beta}$	0.00426	0.00486
St.er./%	4.93	5.41
β_d	0.678	0.790
St.er./%	1.29	1.28
R_{ct}/Ω	16.96	6.944
St.er./%	3.12	2.43
Q_d/SIU	0.00751	0.01511
St.er./%	7.68	7.0
β_l	0.702	0.704
St.er./%	2.85	2.88
R_l/Ω	18.04	6.056
St.er./%	5.41	5.62

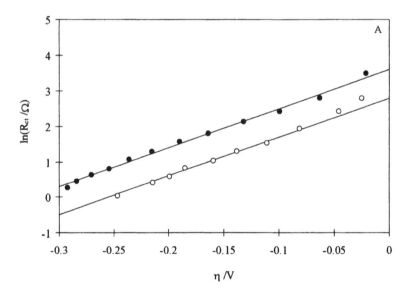

Figure 10. Semi-logarithmic representation of the fitted charge transfer resistance, R_{ct} as a function of the overpotential η. NiMo-B0T deposit (full symbols) and NiMo-B1T deposit (empty symbols). Same conditions as Figure 4.

capacitance but rather, a power-low frequency dependence of capacitance is usually observed. In the equivalent circuit represented in Figure 8, the constant phase elements (CPE) representation of the double layer capacitance C_d and the low frequency capacitance C_l have been included during the fitting procedure. Therefore, the value of C_d for example, can be obtained through the expression of the high frequency time constant described by the high frequency impedance, ZHF which is equal to

$$Z_{HF} = R_s + \frac{R_{ct}}{1 + (j \cdot \omega \cdot R_{ct} \cdot C_d)} \tag{4}$$

where $j^2 = -1$ and β is the CPE exponent.

Therefore, the values of the capacitance C_d, can be obtained from the fitted parameters according to the formula proposed by Brug et al. (J.P. Diard 1988).

$$C_d^\beta = \frac{Q_1}{(R_s^{-1} + R_{ct}^{-1})^{1-\beta}} \tag{5}$$

In (Eq. 5), Q_1 is the double layer CPE element. If $\beta = 1$ hence, Q_1 corresponds to the double layer capacitance C_d.

Using smooth Ni electrode, the usual value of the capacitance C_d is assumed to be ca. $20 \mu F \cdot cm^{-2}$, whereas for HER on Ni-Mo alloys high values of C_d have been reported in various previous works (J.M. Jaksic 2000), (N.V. Krstajic 2008), (M.A. Dominguez-Crespo 2005) and (E. Navarro-Flores 2005). In their papers, the highest values of C_d are essentially attributed to the porosity and roughness of the alloy. From Figure 11, one can see that, the electrode position of Ni-Mo alloy under applied magnetic field leads to high values of the capacitance C_d. The average value is practically potential independent for the NiMo-B1T whereas the minimum value is found for the overpotential values $\eta \leq 0.2V$, for the electrode NiMo-B0T. From this and taking into account the relationship between C_d and the real surface area and the roughness factor (E. Navarro-Flores 2005), we can say that the presence of magnetic field during the alloy electrodeposition process seems to increase the real surface area and the roughness factor of the catalyst. This fact is in agreement with the increase of the exchange current density values obtained through the linear Tafel plots of the stationary polarization curves and may explain the decrease of the electrolytic resistance R_s, which is inversely proportional to the real area of catalyst.

3.5.3 Adsorption parameters

In the literature, generally the admit mechanism for HER in acidic solutions proceeds via three reaction steps, i.e. Volmer-Herovsky reactions which involve metallic hydrogen adsorption-desorption process (Eq. 6 and 7) completed by Tafelchemical desorption (Eq. 8).

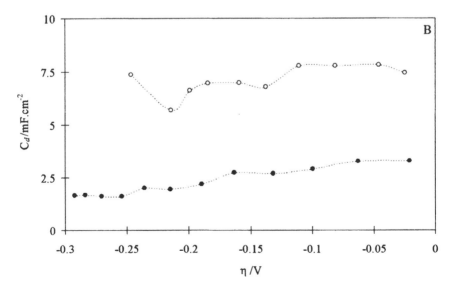

Figure 11. Evolution of the double layer capacitance C_d as function of the overpotential, η. NiMo-B0T deposit (full symbols) and NiMo-B1T deposit (empty symbols). Same conditions as Figure 4.

$$M + H^+e^- = MH_{ad} \qquad (6)$$

$$MH_{ad} + H^+ + e^- = H_2 \qquad (7)$$

$$MH_{ad} + MH_{ad} = 2M + H_2 \qquad (8)$$

As previously reported in (J.M. Jaksic 2000), (L. Birry 2004), (J.P. Diard 1988), the predicted total impedance, Z is composed of two semicircles, located in the positive part of the Nyquist plan. The first semicircle corresponds to the double layer charge process in parallel with the electron charge transfer and the second to the faradic process taking into account the adsorption phenomena. Hence, the corresponding faradic impedance Z_F, may be expressed as

$$Z_F = \frac{1}{R_{ct}} + \frac{1}{Z_{ad}} \qquad (9)$$

where Z_{ad} is the impedance corresponding to the contribution of the adsorbed species involved in reaction mechanism (Eqs. 6–8). According to Diard et al. calculation (J.P. Diard 1988), the corresponding semi-circle may be located in the positive part, when the adsorption contribution is capacitive type, or in the negative part of the Nyquist plane when the adsorption contribution is inductive type. For HER in acidic water, capacitive parameters (CPE element Q_l and R_l) have been introduced in the equivalent circuit (Figure 8) to describe the adsorption relaxation of the impedance in the low frequency range.

Figure 12/13 presents the overpotential evolution of the low frequency resistance R_l and the low frequency C_l for the two deposits. With or without magnetic field the straight lines obtained for R_l data show clearly the exponential nature of this resistance. We obtain here also a decrease of the resistance value when the deposit is obtained under applied magnetic field. However, C_l appears strongly potential dependent and quite constant for $\eta \le -0.15V$ for the two deposits. As for the double layer capacitance, we obtain here also an enhancement of the capacitance values for the NiMo-B1T deposit. This confirms the modification of the real surface area and the morphological structure of the electrodeposited catalyst under applied magnetic field.

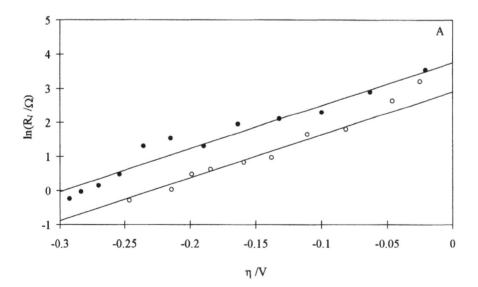

Figure 12. Evolution of the low frequency resistance R_l as function of the overpotential, η. NiMo-B0T deposit (full symbols) and NiMo-B1T deposit (empty symbols). Same conditions as Figure 4.

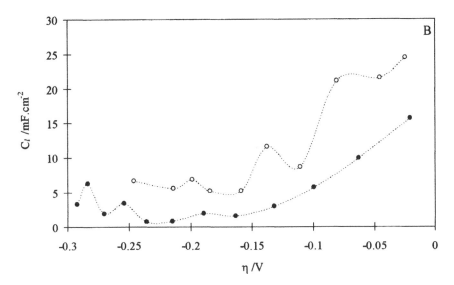

Figure 13. Evolution of the low frequency capacitance C_l as function of the overpotential, η. NiMo-B0T deposit (full symbols) and NiMo-B1T deposit (empty symbols). Same conditions as Figure 4.

4 CONCLUSIONS

In summary, we demonstrate here that a new tool based on the application of an external magnetic field can be used during NiMo alloys electrode position to enhance their catalytic activity for HER.

The comparison of the behavior of the alloy electrode posited with or without magnetic field for HER in acidic solution shows an important enhancement of stationary current densities, a substantial overpotential shift to noble way and a substantial decrease of the ohmic drop in the cell.

Using EIS measurements, the stationary observations have been confirmed and show that, for the NiMo catalyst electrodeposited under applied magnetic field, high double layer capacitance values; low charge transfer resistance and low polarisation resistance have been obtained. These observations indicate clearly, the effect of external magnetic on the real surface area and the morphology structure of the catalyst. This work is expected to have important contribution to improve the practical ways to obtain hydrogen from water electrolysis. Work is in progress to optimize the magnetic field and the bath deposition conditions to optimize the experimental conditions to get substantial alloy activity modifications with low coast.

REFERENCES

Aaboubi, O., Chopart, J.P., J.D.-A.O.C.G.B.T. (1990). Magnetic field effects on mass transport. *J Electrochem Soc 137*, 1796–1804.

Aaboubi, O., J.C. (2010). Magnetic field effects on molybdenum based alloys. *ECS Trans 25*, 27–34.

Aogaki, R., Fueki, K., T.M. (1975). Application of magnetohydrodynamic effects to the analysis of electrochemical reactions. *Denki Kagaku 43*, 504–514.

Birry, L., A.L. (2004). Studies of the hydrogen evolution reaction on raney nickelmolybdenum electrodes. *J. Appl. Electrochem. 34*, 735–749.

Bockris, J.O.M., S.K. (Ed.) (1993). *Surface electrochemistry: a molecular level approach*, New York. Plenum Press.

Brug, G.J., Ven den Eeden, A.L.G., M.S.-R.J.S. (1984). The analysis of electrode impedances complicated by the presence of a constant phase element. *J Electroanal Chem 176*, 275–295.

Devos, O., Olivier, A., J.C.-O.A.G.M. (1998). Magnetic field effects on nickel electrodeposition. *J Electrochem Soc 145*, 401–405.

Diard, J.P., Landaud, P., B.L.G.C.M. (1988). Calculation, simulation and interpretation of electrochemical impedance. part ii. interpretation of volmer-heyrovsky impedance diagrams. *J Electroanal Chem 255*, 1–20.

Dincer, I. (2002). Technical environmental and exergetic aspects of hydrogen energy systems. *Int J Hydrogen Energy 27*, 265–285.

Dominguez-Crespo, M.A., Plata-Torres, M., A.M.T.-H.E.A.-E.J.H.-L. (2005). Study of hydrogen evolution reaction on $Ni_{30}Mo_{70}$, $Co_{30}Mo_{70}$, $Co_{30}Ni_{70}$ and $Co_{10}Ni_{20}Mo_{70}$ alloy electrodes. *Mat Charac 55*, 8391.

Dunn, S. (2002). Hydrogen futures: toward a sustainable energy system. *Int J Hydrogen Energy 27*, 235–264.

Fahidy, T. (1983). Magnetoelectrolysis. *J Appl Electrochem 13*, 553–563.

Fahidy, T. (2001). Characteristics of surfaces produced via magnetoelectrolytic deposition. *Prog Surf Sci 68*, 155–188.

Han, Q., Cui, S., N.P.-K.L.X.A.W. (2010). Study on pulse plating amorphous ni-mo alloy coating used as her cathode in alkaline medium. *Int J Hydrogen Energy 27*, 5194–5201.

Hinds, G., Coey, J.M.D., M.L. (2001). Influence of magnetic forces on electrochemical mass transport. *Electrochem Com 3*, 215–218.

Hugo, A., Rutter, P., S.P.A.A.G.Z. (2005). Hydrogen infrastructure strategic planning using multi-objective optimization. *Int J Hydrogen Energy 30*, 1523–1534.

Jaksic, J.M., Vojnovic, M.V., N.K. (2000). Kinetic analysis of hydrogen evolution at nimo alloy electrodes. *Electrochim Acta 45*, 4151–4158.

Khotari, R., Buddhi, D., R.S. (2005). Studies on the effect of temperature of the electrolytes on the rate of production of hydrogen. *Int J Hydrogen Energy 30*, 261–263.

Krstajic, N.V., Jovic, V.D., L.G.-K.B.J.A.A.G.M. (2008). Electrodeposition of nimo alloy coatings and their characterization as cathodes for hydrogen evolution in sodium hydroxide solution. *Int J Hydrogen Energy 33*, 3676–3687.

Lee, J., Gao, X., L.H.H.W. (1995). Influence of magnetic field on the voltametric response of microelectrodes in highly concentrated organic redox solutions. *J Electrochem Soc 142*, L90–L92.

Leventis, N., Chen, M., X.G.M.C.P.Z. (1998). Electrochemistry with stationary disk and ring-disk millielectrodes in magnetic fields. *J Phys Chem B 102*, 3512–3522.

Martinez, S., Metikos-Hukovic, M., L.V. (2006). Electrocatalytic properties of electrodeposited ni15mo cathodes for the her in acid solutions: Synergistic electronic effect. *J Mol Catal A Chem 245*, 114–121.

Msellak, K., Chopart, J.P., O.J.O.A.J.A. (2004). Magnetic field effects on ni-fe alloys codeposition. *J Mag Mag Mat 281*, 295–304.

Navarro-Flores, E., Chong, Z., S.O. (2005). Charaterization of ni, ni-mo, niw and nife electroactive coatings as electrocatalysts for hydrogen evolution in acidic medium. *J Mol Cata A Chem 226*, 179–197.

Rabah, K.L., Chopart, J.-P., H.S.S.S.O.A.M.U.D.E.-J.A. (2004). Analysis of the magnetic force effect on paramagnetic species. *J Electroanal Chem 571*, 85–91.

Russell, H.J., G.T. (2007). An overview of materials for the hydrogen economy. *JOM 59*, 50–55.

Zabinski, P.R., Jarek, A., R.K. (2009). Effect of applied external magnetic field on electrodeposition of cobalt alloys for hydrogen evolution in 8m naoh. *Magnetohydrodynamics 45*, 275–280.

Lecture Notes on Impedance Spectroscopy – Kanoun (ed)
© 2012 Taylor & Francis Group, London, ISBN 978-0-415-69838-2

Scalable impedance spectroscopy: Comparative study of sinusoidal and rectangular chirp excitations

Mart Min
Department of Electronics, Tallinn University of Technology, Tallinn, Estonia
Institut für Bioprozess- und Analysenmesstechnik, Heilbad Heiligenstadt, Germany

Toomas Parve
Department of Electronics, Tallinn University of Technology, Tallinn, Estonia

Raul Land
Department of Electronics, Tallinn University of Technology, Tallinn, Estonia
Institut für Bioprozess- und Analysenmesstechnik, Heilbad Heiligenstadt, Germany

Paul Annus
Department of Electronics, Tallinn University of Technology, Tallinn, Estonia
ELIKO Competence Centre, Tallinn, Estonia

ABSTRACT: Wideband excitation enables to cover a wide frequency range during a short time interval. As a result, the measurement system can provide spectral information matched with the dynamics of the impedance under study. The chirp signal has a unique property: bandwidth of its power spectrum can remain constant at different durations of chirp pulses. This quality enables to perform scalable spectroscopy. The paper describes and analyses the properties of two waveforms of chirp excitation: the sine wave and rectangular wave chirps.

Keywords: impedance spectroscopy, wideband excitation, chirp excitation, scalability, sinusoidal chirp, rectangular chirp

1 INTRODUCTION

Under the term scalability in impedance spectroscopy we consider the possibility to scale the excitation signal so that its duration matches the dynamics of an object under study with variable impedance via covering a required frequency range (bandwidth B) independently on the duration T of excitation (Nahvi and Hoyle 2008), (Nahvi and Hoyle 2009), (Hoyle and Nahvi 2008), (M. Min and Land 2008).

In principle, such the scalability is valid for sine wave chirp signals in both time and frequency domain (double scalability). The chirps can have very short duration and contain only a fraction of one cycle, but still cover the required frequency range, though the spectral density becomes low (Min and Annus 2010). Chirps are not periodic signals, they do not have certain time periods as a sine wave function, under the cycle we mean a phase rotation by 2π, see equation (1).

The sine wave chirp is not the simplest signal to generate. Rectangular waveforms are much simpler to generate by the aid of digital electronics and are of greater interest, therefore (M. Min and Land 2009).

In Fig. 1, there is given a generalized architecture of the measurement system for performing the scalable impedance spectroscopy. Before measurement, we can separately determine the both, duration T from t_1 to t_2 and frequency band B from f_1 to f_2 of spectroscopy, and change these values independently to match the measurement with the speed of changing of the complex impedance under study $Z(f,t)$.

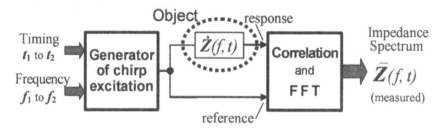

Figure 1.   Basic architecture of a system for performing scalable impedance spectroscopy.

2   SINUSOIDAL CHIRP

An outstanding property of chirp functions is the possibility to set useful excitation bandwidth B independently on duration of the chirp pulse T. This can be done by choosing an appropriate rate B/T, if the chirping is linear. Such the linear chirp pulses (Fig. 2) can be described as

$$ch(t) = A\sin\left[2\pi\frac{B}{T}\frac{t^2}{2}\right]$$

(1)

where $0 \le t \le T$.

Scalability of the sinusoidal chirps in time domain means that if we change the duration T of a linear chirp, we have to change also the speed of frequency changing (or chirping), thus maintaining the frequency range or spectral bandwidth $B = f_2 - f_1$ constant, see the chirp signals with different durations $T = 250\mu s$ (Fig. 2a) and $T = 1000\mu s$ (Fig. 2b). If scaled properly, the changes in the pulse duration T reflect only in spectral density, but not in bandwidth B (Fig. 3).

Scalability of the chirp signals in frequency domain means that we can change the bandwidth B of linear chirp maintaining the pulse duration T constant (Fig. 4). This becomes possible through changing the speed B/T of chirping. The changes in bandwidth reflect in spectral density of chirp signals, but not in bandwidth and almost not in the shape of spectral density function (Fig. 5). The sine wave has a crest factor $CF = A/RMS = 2^{1/2} = 1.414$. Therefore, the root-mean-square value (RMS) of the sinusoidal chirp with unity voltage amplitude $A = 1V$ is equal to $2 - 1/2 = 0.707V$, and the density of energy distribution in time, or power P as an amount of energy per time unit, is equal to $0.5V^2/s$.

RMS spectral densities of two different duration chirps in Fig. 2a and Fig. 2b are given in Fig. 3a and Fig. 3b, wherein the both chirps have the same bandwidth $B = 100kHz$. Spectral densities differ two times if the durations differ four times: $35\mu V/Hz^{1/2}$ for the chirp with duration $T = 250\mu s$ (Fig. 2a), see Fig. 3a, and $70\mu V/Hz^{1/2}$, if the duration $T = 1000\mu s$ (Fig. 2b), see Fig. 3b.

Spectral densities of two chirps with the same duration $T = 250\mu s$ in Fig. 4a and Fig. 4b differ also two times due to the four times difference in bandwidth B: $35\mu V/Hz^{1/2}$ in the case of $B = 100kHz$ (Fig. 5a), and $17\mu V/Hz^{1/2}$, if the bandwidth $B = 400kHz$ (Fig. 5b). Excitation energy E characterizes the excitation level and depends proportionally on duration of the excitation pulse $E = P \times T$. Therefore, it is reasonable to use longer excitation pulses for obtaining better signal-to-noise ratio, but there are limitations. The main limiting factor is the speed of impedance variations (dynamics): current state of the object under study must remain unchanged (at least conditionally) during the excitation time T.

Thanks to specific properties of the chirp function, it possible to match the needs for bandwidth, time, signal-to-noise ratio and dynamic requirements.

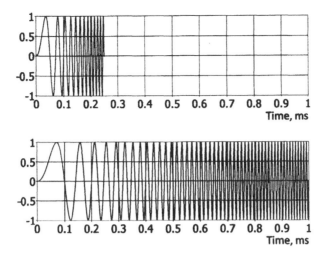

Figure 2. Excitation chirp pulses with normalised amplitude $A = 1V$ and different durations: (a)-$T = 250\mu s$ containing 12 signal cycles, and (b)-$T = 1000\mu s$ and contains 48 cycles; both signals are scaled into the spectral bandwidth $B = 100kHz$.

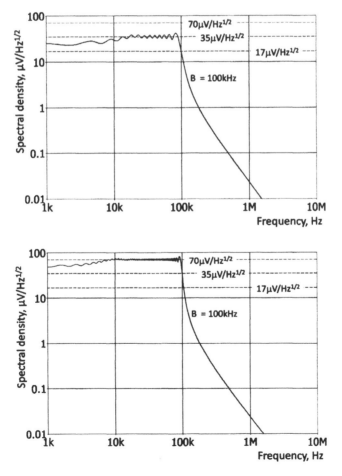

Figure 3. RMS spectral density functions of two chirp pulses described in Fig. 2 with durations: (a)-$T = 250\mu s$, and (b)-$T = 1000\mu s$, both with $B = 100kHz$.

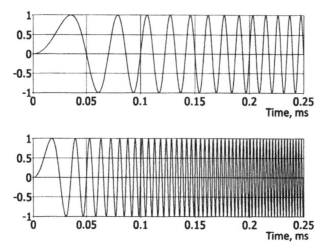

Figure 4. Excitation chirp pulses with the same normalised amplitude $A = 1V$ and duration $T = 250\mu s$, but different bandwidths: (a)-$B = 100kHz$ (contains 12 cycles), and (b)-$B = 400kHz$ (contains 48 cycles).

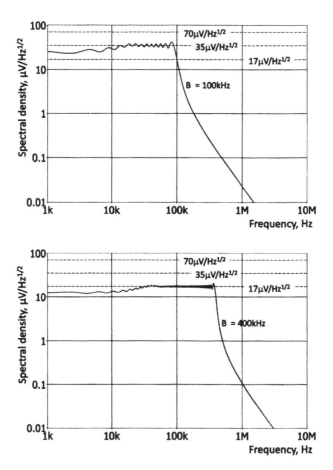

Figure 5. RMS spectral density functions of two chirp pulses with the same duration $T = 250\mu s$ (see Fig. 4), but different bandwidth values: (a)-$B = 100kHz$, and (b)-$B = 400kHz$.

36

3 RECTANGULAR CHIRPS

Rectangular chirps are depicted in Fig. 6: first a 2-state or binary signum-chirp without any zero level states (Fig. 6a), and second, a 3-state ternary chirp with zero level states as in Fig. 6b. The simple signum-chirp can be expressed on the bases of a sine wave chirp (1) as follows:

$$Sign\{ch(t)\} = sign\left\{A \cdot \sin\left[2\pi\frac{B}{T} \cdot \frac{t^2}{2}\right]\right\} \qquad (2)$$

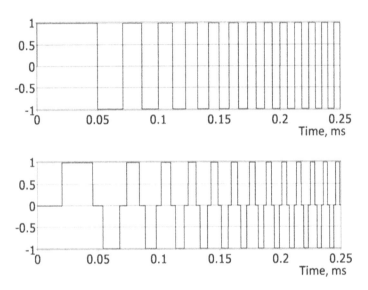

Figure 6. Rectangular chirps: (a)-binary (2-state) or signum-chirp pulse, and (b)-ternary (3-state) rectangular chirp pulse.

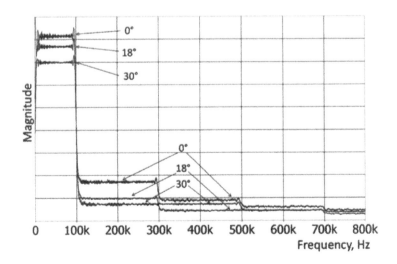

Figure 7. RMS spectra of rectangular chirp pulses: of a binary chirp without zero state (0°), and of ternary chirps with zero states during 18° and 30°.

37

The crest factor of binary chirps has a unity value, $CF = 1$. That is, its peak value V_P and *RMS* value are equal. For $V_P = 1V$, also the *RMS* value and power P level will have the unity value. It means that by energy the signum-chirp is two times more powerful than the sinusoidal one. This is an important benefit besides simplicity to generate in practical devices.

Unfortunately, not all the generated energy falls into the useful frequency range, see the bandwidth $B = 100 kHz$ in Fig. 7. However, analysis shows that the lost energy is not big— about 85% of energy becomes useful when the binary signum-chirp is generated. Somewhat more, about 90% of generated energy becomes useful in the case of 3-state ternary chirps when duration of the zero state is 18°, and even slightly more, if the zero state extends to 30° (Min and Annus 2010). The most practical solution could be 22.5° or 1/16th of the full 360° cycle, because of simplicity to generate using digital hardware.

4 CONCLUSION

Chirps are very good candidates for using as excitation signals in scalable impedance spectroscopy if the impedance under study is dynamic, that is, changes in time.

Sine wave chirps are flexible, they can be used in very short time spectroscopy—only fractions of one cycle can cover the required measurement bandwidth (Min and Annus 2010). Also near to 100% of the generated energy falls into the measurement bandwidth B. But the crest factor 1.414 shows that only half of signal energy $E = 0.5V^2 \times T$ within a limited range of amplitudes between $+V$ and $-V$ is available.

Rectangular chirps are easy to generate by the aid of nowadays digital electronics, their crest factor is low. For binary chirps, $CF = 1$, and the generated energy is $E = V^2 \times T$, but a part this of energy falls outside the measurement bandwidth B. Fortunately, this part is small, only about 15% for binary chirps, and even less in the case of 3-level ternary chirps. The rectangular chirps are energetically more effective than sinusoidal ones, only some application restrictions exist due to the complicated harmonic content of rectangular signals. This is a subject of further studies.

ACKNOWLEDGMENT

The authors thank colleagues from the EU FP6 Marie Curie ToK project No. 29857 InFluEMP team at the Institut für Bioprozess- und Analysenmesstechnik in Heilbad Heiligenstadt, Germany. Special thanks belong to Dr. Uwe Pliquett for his guiding discussions on the principles of impedance spectroscopy. The research was supported by European Regional Development Fund and Enterprise Estonia through the Competence Centre ELIKO.

REFERENCES

Hoyle, B.S. and Nahvi, M. (2008). Spectro-tomography an electrical sensing method for integrated estimation of component identification and distribution mapping in industrial processes. *Proc. 2008 IEEE Sensor Conference*, pp. 807–810.

Min, Land, R., T.P.T.P. and Annus, P. (2010). Broadband spectroscopy of a dynamic impedance. *Journal of Physics: Conference Series 224*.

Min, M.T. Paavle, P.A. and Land, R. (2009). Rectangular wave excitation in wideband bioimpedance spectroscopy. *Proc. 4th International Workshop on Medical Measurements and Applications*, pp. 268–271.

Min, M.U. Pliquett, T.N.A.B.P.A. and Land, R. (2008). Broadband excitation for short-time impedance spectroscopy. *Physiol. Meas. 29*, 185–192.

Nahvi, M. and Hoyle, B.S. (2008). Wideband electrical impedance tomography. *Meas. Sci. Technol. 19*, pp. 9.

Nahvi, M. and Hoyle, B.S. (2009). Electrical impedance spectroscopy sensing for industrial processes. *IEEE Sensors Journal Vol. 9, No. 12*, pp. 1808–1816.

Lecture Notes on Impedance Spectroscopy – Kanoun (ed)
© *2012 Taylor & Francis Group, London, ISBN 978-0-415-69838-2*

Harmonic controlled excitation in galvanostatic impedance spectroscopy

M. Lelie, M. Kiel & D.U. Sauer
Electrochemical Energy Storage Systems, Institute for Power Electronics and Electrical Drives (ISEA), RWTH Aachen University, Aachen, Germany

ABSTRACT: Impedance Spectroscopy has become a significant tool not only in corrosion and coatings research, but also in material science, biomedical engineering as well as fuel cell and battery research. For the latter, Electrochemical Impedance Spectroscopy (EIS) is of increasing importance for modelling and diagnosis. Special care must be taken in order to determine parameters of those nonlinear devices, since there is a difference between large and small signal impedance (J. Kowal 2009).

Therefore, the nonlinear behaviour of fuel cells and batteries may lead to false measurement results, if the measurement is not performed thoroughly enough. In order to stay in a quasi-linear measurement region, it is recommended to maintain a voltage response of the device under test not greater than $10mV$ (E. Barsoukov 2005). However, in (M. Kiel 2008), a different method for maintaining a quasi-linear state is suggested. It analyses higher harmonics in the response signal, in order to control the excitation amplitude.

This paper explains the correlation between nonlinearities and higher harmonics. Afterwards it shows the theory of a new controlling algorithm for the excitation amplitude based on the measurement of occurring harmonics. The working principle of the algorithm is explained on the basis of simulation results. Measurement results will show the differences between pure galvanostatic excitation, voltage controlled excitation, and harmonic controlled excitation.

Concluding, the new algorithm's performance and it's limitations in the practical implementation are analysed.

Keywords: impedance spectroscopy; harmonics; batteries; THD

1 INTRODUCTION/MOTIVATION

Galvanostatic impedance spectroscopy on batteries has to cope with the nonlinearity of the current-voltage relationship of the device under test.

If the excitation amplitude in nonlinear systems is too high, the response signal becomes distorted. Figure 1 illustrates this effect in time and frequency domain. The excitation signal is purely sinusoidal and therefore appears as a single peak in the frequency domain. The response signal from the nonlinear transfer function however is deformed and its representation in the frequency domain shows higher harmonics as well as a DC offset.

In addition, as shown in (M. Kiel 2008) and (Darowicki 1995), the value of the response fundamental frequency is influenced by the distortion due to the nonlinearity of the transfer function, leading to an erroneous measurement of the fundamental frequency impedance. The resulting error in relation to the ideal small signal excitation depends on the fundamental frequency as well as the excitation amplitude. The occurrence of harmonics as well depends on the frequency, when using a constant excitation amplitude (M. Kiel 2008), (Darowicki 1995).

In order to stay in a quasi-linear state during the measurement, literature suggests maintaining a maximum voltage response of $10mV$ around the operation point (E. Barsoukov 2005).

Figure 2 shows the impedance spectra of a $2V$, $5Ah$ lead acid battery cell, which is recorded using three different excitation amplitudes. Each recording is made directly one after another

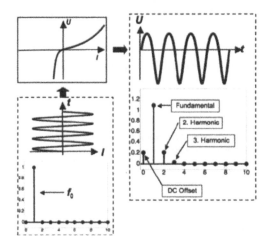

Figure 1. Effect of a nonlinear transfer function on a sinusoidal excitation signal. If the excitation amplitude is too high, the frequency analysis reveals a significant occurrence of higher harmonics as well as a DC offset.

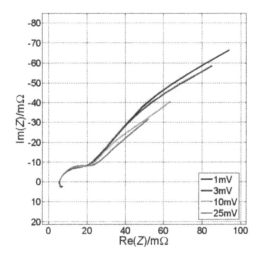

Figure 2. Impedance spectra of a $2V$, $5Ah$ lead acid battery cell. The four spectra are recorded directly one after another without changes in temperature or State of Charge (which is 80%). The excitation amplitude is changed between the spectra from $1mV$ to $25mV$.

without changes in temperature or SOC. It can be seen, that even with an excitation amplitude of less than $10mV$ the results of the spectra differ significantly.

In (M. Kiel 2008) it is suggested to use the analysis of the resulting harmonics in order to control the excitation amplitude. The idea is that in a first step a value directly related to the characteristic of the nonlinearity is calculated by analysing the occurring harmonics. Then this value is used to control the excitation in order to achieve a quasi-linear state for the impedance measurement.

2 THEORY AND SIMULATION

In a first step, a MatLab Simulink based simulation is used for testing the new excitation control algorithm.

A simplified Randles circuit model as shown in Figure 3 is implemented and tested with three different excitation control algorithms.

The model consists of a pure ohmic series resistance R_s, a capacitance C_{dl} representing the double layer capacity and a nonlinear resistance R_{ct}, following the Butler-Volmer relationship.

The three simulated algorithms for controlling the excitation amplitude are

- Constant current excitation
- Voltage controlled excitation
- Harmonic content controlled excitation.

The first control algorithm is a simple galvanostatic excitation without any control of the excitation amplitude. The second method consists of a galvanostatic operation with a superimposed voltage regulation algorithm (hybrid potentio- galvanostatic operation). The third excitation control variant takes into account the occurring harmonics in relation to the fundamental frequency. For this harmonic content controlled algorithm, a figure of merit for the harmonic content is introduced:

$$THD^* = 100 \cdot \frac{\sqrt{\hat{U}^2_{2f_0} + \hat{U}^2_{3f_0}}}{\sqrt{\hat{U}^2_{f_0}}} \tag{1}$$

$\hat{U}^2_{f_0}$ is the absolute value of the fundamental frequency; $\hat{U}^2_{2f_0}$ and $\hat{U}^2_{3f_0}$ are the absolute values of the second and the third harmonic frequencies.

The definition of THD^* is related to the definition of the Total Harmonic Distortion (THD), which is defined as:

$$THD_\% = 100 \cdot \frac{\sqrt{\sum_{i=2}^{n} U_i^2}}{U_1} \tag{2}$$

where U_1 is the absolute value of the fundamental frequency and $U_2, U_3, ..., U_n$ are the amplitudes of the harmonic frequencies accordingly. THD^* is introduced, because it could be easily implemented into the measurement device since it requires less computational power to only calculate the second and third harmonic.

The usage of only the first two harmonics increases the performance of the algorithm compared to a method where more harmonics are used. Anyway, as shown in (M. Kiel 2008), these are the most relevant ones, as the higher harmonics are significantly decreasing in magnitude This can also be shown in a measurement (see Figure 4). For the simulation, the excitation amplitude was controlled to achieve a THD^* value of 0.5%.

For every control algorithm, the results are compared to results from the simulation using a model with a linear resistance R_{ct}.

Therefore, a resulting error due to the non-ideal small signal excitation can be calculated for each control algorithm.

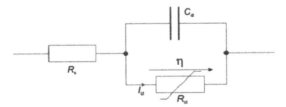

Figure 3. Simplified Randles circuit with series resistance R_s, double layer capacity C_{dl}, and nonlinear charge transfer resistance R_{ct}.

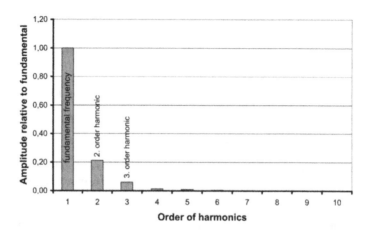

Figure 4. Harmonics of the voltage response of a single lead acid cell with $5Ah$ nominal capacity at a measurement frequency of $0.562Hz$. At this frequency the harmonic content's amplitudes show a maximum with the used battery in a frequency range between $10kHz$ and $10mHz$. The measurement was taken with a Zahner Zennium Electrochemical workstation. It can be seen, that the second and the third harmonics are the most relevant ones.

Figure 5 shows the results of the simulation. For the constant current excitation the error with respect to the linear model as well as the harmonic content THD^* increase, as expected, with decreasing frequency. The current amplitude is constant over the complete frequency range and the voltage response follows the impedance characteristic of the circuit.

For the response signal excitation control the error with respect to the linear model increases, but reaches a much smaller value than in case of constant current excitation. The excitation signal is also much smaller compared to constant current excitation and the response signal remains constant over the complete frequency range as with voltage controlled excitation. The measurement results are therefore considered to be the same as with response signal controlled excitation. The harmonic content reaches the same value as with response controlled excitation, but reaches the maximum value of 0.5% at higher frequencies compared with response signal controlled excitation. The excitation current remains the same as with constant current excitation and drops rapidly, when the harmonic content reaches its maximum value of 0.5%. The voltage response in the high frequency range is also the same as with constant current excitation, which is much higher compared to the response signal controlled excitation. For practical considerations this is a big advantage, because the smallest impedances of the spectrum occur at high frequencies. A high excitation current leads to a higher voltage response and therefore to a better signal to noise ratio.

2.1 *Measurement device*

In order to confirm the theoretical simulations on a real measurement with real batteries, an impedance measurement device was modified to be able to use the new controlling algorithm.

The device "EISmeter", produced by Digatron Firing Circuits, was developed in cooperation between ISEA of RWTH Aachen University and Digatron. The device normally operates in galvanostatic mode, with response signal controlled excitation.

The excitation current is controlled during the measurement, relatively to the voltage response of the battery under test and a pre-given value called U_ideal. Usually, U_ideal is set to 3 or $10mV$ per cell by the user. The frequency range of the device reaches from $12kHz$ down to a few μHz.

Due to the internal construction, the measurement of THD^* is only possible below a measurement frequency of $1kHz$. However, for most batteries, the occurrence of nonlinearities leading to harmonics in the response signal are considered at much lower frequencies and therefore the $1kHz$ boundary is sufficient.

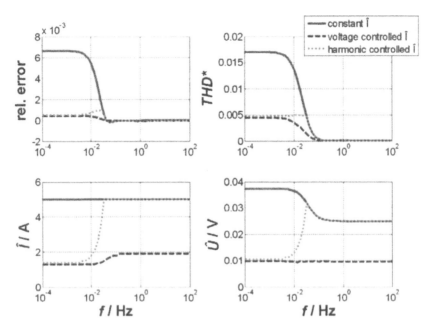

Figure 5. Simulation results of three different amplitude control algorithms: constant current amplitude, response (voltage) signal controlled excitation amplitude and harmonic content (THD^*) controlled excitation amplitude.

2.2 Excitation control

In order to verify the simulations by comparing with real measurement results, the THD^* control algorithm was implemented into the EIS meter.

While the simulation adjusts the excitation amplitude for every frequency, starting each time with the same maximum amplitude, the implementation on the EIS meter is optimised to save measurement time. Every frequency is measured with a given amplitude at first. The excitation amplitude is then adjusted according to the THD^* value, only if the measured THD^* differs ±5% from the pre-given value. In this case, a linear relationship between THD^* and the excitation amplitude is assumed. If the THD^* value does not fit after this linear adjustment, the excitation amplitude is adjusted again with a fixed step size, until the THD^* value is within the desired limits.

2.3 Measurement results

In the following, measurement results with the THD^* excitation control algorithm are shown in comparison with response signal excitation control (hybrid potentio—galvanostatic control). A small $12V$, $1.1Ah$ lead acid battery and a $2.5Ah$ $LiFePO_4$ battery are measured.

The results shown in Figure 7 correlate with the results of the simulation: The current amplitude and therefore the voltage response in the high frequency range is much higher for the harmonic controlled excitation compared to the two response signal controlled excitation curves. In the lower frequency range, the excitation amplitude drops rapidly in order to fulfil the $THD^* = 0.5\%$ criterion. It can also be seen, that the harmonic content THD^* is not always maintained at 0.5%. This is due to noise considerations; the measurement device sets a minimum excitation amplitude of $1mA$. However, the harmonic content of the THD^* controlled measurement in the low frequency range is lower than with the response signal controlled measurements. This indicates that a better quasi-linear state was achieved in this frequency range. In the frequency range between $1kHz$ and $500Hz$ the harmonic content is higher compared to the voltage controlled measurement, but the bode plot of the measurements in Figure 6 shows a good correlation of all measurements in this frequency range.

43

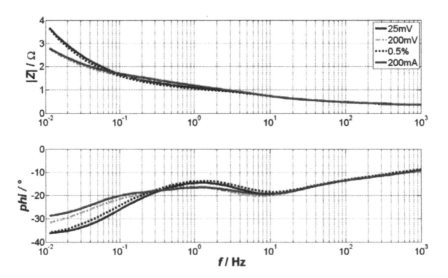

Figure 6. Measurement result on a $12V$, $1.1Ah$ lead acid battery. The spectra are recorded directly one after another and at constant temperature and SOC. The figure represents the results of a measurement with response signal controlled current excitation and a voltage limit of $25mV$ (i.e. $\approx 4mV$/cell) and $200mV$ (i.e. $33mV$/cell). Also the results of a measurement with a constant current amplitude of $200mA$ and measurement with a THD^* controlled excitation current amplitude with a maximum THD^* of 0.5% is shown.

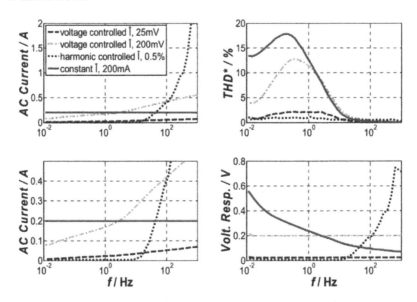

Figure 7. Measurement results of a $12V$, $1.1Ah$ battery for four different control strategies: Response signal controlled excitation (25 and $200mV$), constant current amplitude excitation ($200mA$) and harmonic content (THD^*) controlled excitation (0.5%).

The measurement results of the $3.2V$ $LiFePO_4$ battery are shown in Figure 8. The upper left diagram shows the characteristic of the impedance modulus. It can be seen that all algorithms lead to the same results. Again, the THD^* controlled measurement provides higher excitation and response signals, if the distortion due to nonlinearities remains in the pregiven range. It can be seen, that it is difficult to maintain the 0.5% THD^* criteria taking into account measurement noise because of a given current limit.

44

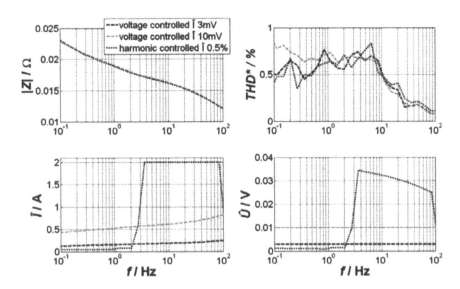

Figure 8. Measurement results of a $2.5Ah$ $LiFePO_4$ $3.2V$ battery. Two measurements with response signal controlled amplitude were performed and compared with a THD^* controlled measurement.

2.4 *Limitations*

As shown in the measurement results, the harmonic-content-based controlling algorithm suffers from a few limitations.

As already seen above, it is not always possible to maintain a given THD^* limit during the whole measurement. The black dotted curve in Figure 9 shows the THD^* values measured during a spectrum on a $12V$, $1.3Ah$ lead acid battery ($22°C$, SOC 80%).

As it is not possible to reduce the amplitude to arbitrary low values, because of noise problems (see below), a lower limit for the current amplitude and a lower limit for the minimal necessary voltage response has to be implemented.

The impact of noise on the THD^* value becomes visible in Figure 10. In the right part of the figure, the THD^* value for a $5Ah$, $2V$ lead acid battery in dependency on the excitation current is shown. It can be seen, that THD^* decreases with decreasing current, but increases again, if the current is too low ($5mA$ curve). Therefore the lower limit for the excitation current has to be chosen in a way to find a compromise between the point where the noise begins to interfere with the THD measurement and the point where the harmonics don't decrease further when decreasing the amplitude.

Measurements were taken to determine the noise border of the measurement device. Based on these measurements the voltage response limit could be defined as $1mV$/cell for this particular measurement device.

This measurement identifies another drawback of the THD^* controlled measurement: $1mV$/cell can be resolved easily by the device in an U_ideal -controlled measurement, but since the occurring harmonics decrease rapidly in amplitude with increasing order, the resolution of the higher harmonics gets worse using the THD^* controlled measurement.

Another problem is the extremely non-linear relation between the excitation amplitude and the occurring harmonics. In contrary e.g. to the U_ideal control it is not possible to simply calculate the necessary current amplitude for one frequency based on a measurement with an arbitrary amplitude at the same frequency. Thus depending on the quality of non-linearity it sometimes can be necessary to measure the same frequency two or three times in order to finally meet either the given THD^* limit, or one of the lower boundaries for the current amplitude or the necessary voltage response.

Furthermore the described relationship differs for each tested device. The plots in Figure 10 show the THD^* data, which are measured during several impedance spectra with

45

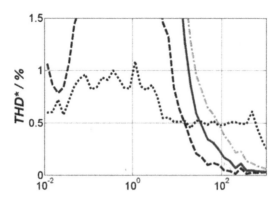

Figure 9.   Enlarged version of the THD^* plot from Figure 7.

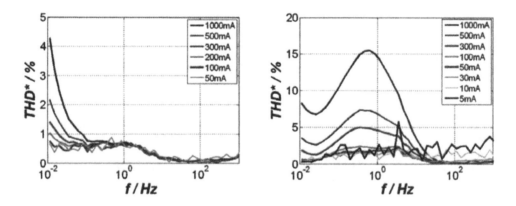

Figure 10. THD^* over frequency for a 2.5Ah $LiFePO_4$ cell (left) and a 5Ah lead acid battery (right) and for different constant currents.

different, constant current amplitudes. These measurements are carried out on a single $LiFePO_4$ cell (26650, 2.5Ah, 80% SOC) and on a single lead acid cell (5Ah, 80% SOC). From the plotted values it can be seen, that the lead acid battery creates much higher harmonics in the whole tested frequency range. The $LiFePO_4$ cell however shows nearly no dependency between the measured THD^* and the current amplitude down to about 100mHz the resolution of the higher harmonics gets worse using the THD^* controlled measurement. Another problem is the extremely non-linear relation between the excitation amplitude and the occurring harmonics. In contrary e.g. to the $U_i deal$ control it is not possible to simply calculate the necessary current amplitude for one frequency based on a measurement with an arbitrary amplitude at the same frequency. Thus depending on the quality of non-linearity it sometimes can be necessary to measure the same frequency two or three times in order to finally meet either the given THD^* limit, or one of the lower boundaries for the current amplitude or the necessary voltage response. Furthermore the described relationship differs for each tested device. The plots in Figure 10 show the THD^* data, which are measured during several impedance spectra with different, constant current amplitudes. These measurements are carried out on a single $LiFePO_4$ cell (26650, 2.5Ah, 80% SOC) and on a single lead acid cell (5Ah, 80% SOC). From the plotted values it can be seen, that the lead acid battery creates much higher harmonics in the whole tested frequency range. The $LiFePO_4$ cell however shows nearly no dependency between the measured THD^* and the current amplitude down to about 100mHz. These variations are a major problem, when trying to find a proper controlling algorithm for the current amplitude based on the THD^* values. In the first attempts a true

46

linear approach, similar to the one used in the $U_i deal$ control was tested with unsatisfying results. So the next approach takes into account using a "brute-force" method, which iteratively adapts the current amplitude until the wanted THD^* limit is met (similar to the method used during the simulation). Using this procedure the THD^* limits are fulfilled perfectly. In a practical application however, the negative aspect of this method is the huge amount of time needed for each single frequency step.

To decrease the measurement time, the "brute-force" method is modified, calculating an approximate value for the needed current amplitude based on two measurements with different amplitudes. Using these two amplitude/THD^* pairs, a linear approximation is calculated. Afterwards, only small additional corrections are necessary to finally reach the given THD^* limit.

Nevertheless, in general a THD^* controlled measurement takes longer than e.g. a simple $U_i deal$ controlled, or a pure galvanostatic measurement.

3 CONCLUSION/OUTLOOK

The work described in this paper shows that it is possible to use the information about the occurring harmonic content to control the excitation amplitude. A simulative approach was chosen to compare the new algorithm to two well known and widely used controlling algorithms. In doing so, a measure for the harmonic content was defined based on the fact, that in the given application only the first two harmonics are relevant.

Afterwards the simulated algorithm was implemented into the Digatron Impedance Spectroscope, in order to perform measurements on real batteries. The measurement results show, that the procedure works in principle and that the main goal of an increased SNR could be achieved. However the practical implementation also shows the algorithms limitations when used in combination with real electrochemical systems.

The implemented solution works well, but leads to an unsatisfactory long measurement time for some devices, since the relationship between the harmonic content and the excitation amplitude varies with each device under test and can not be predicted. In cases where the improved SNR typically would not be necessary, this increased time is a major disadvantage. However, in contrast to the $U_i deal$ parameter, a great advantage of the THD^* method is the parameter's independence of the DUT (e.g. number of cells in a battery).

To overcome this problem without loosing the advantages of the THD controlled procedure, another approach would be using a combined controlling method. By using THD^* control as long as the specified $U_i deal$ is exceeded and afterwards using the voltage controlled excitation (while measuring from high to low frequencies), the advantages of both methods could be combined.

ACKNOWLEDGEMENTS

This work was kindly financed by the E.ON International Research Initiative within the project BEST.

REFERENCES

Barsoukov, E., J.M. (2005). *Impedance Spectroscopy*. New York: Wiley & Sons.
Darowicki, K. (1995). The amplitude analysis of impedance spectra. *Electrochimica Acta 40*, 439–445.
Kiel, M., Bohlen, O., D.U.S. (2008). Harmonic analysis for identification of nonlinearities in impedance spectroscopy. *Electrochimica Acta 53*, 7367–7374.
Kowal, J., Hente, D., D.U.S. (2009). Model parameterization of nonlinear devices using impedance spectroscopy. *IEEE Trans. Instrumentation and Measurement 58*, 2343–2350.

Lecture Notes on Impedance Spectroscopy – Kanoun (ed)
© 2012 Taylor & Francis Group, London, ISBN 978-0-415-69838-2

Hardware concept and feature extraction for low-cost impedance spectroscopy for semiconductor gas sensors

Peter Reimann, Marco Schler, Simon Darsch, Tobias Gillen & Andreas Schtze
Laboratory for Measurement Technology, Department of Mechatronics, Saarland University, Saarbrücken, Germany

ABSTRACT: Impedance spectroscopy (IS) is suitable for different applications like fuel cells, batteries or food analysis. Another field of interest with promising results are IS measurements in semiconductor (SC) gas sensor applications. The high costs of measurement systems impede the use of IS in field tests, which are essential for developing sensor systems efficiently. In this paper we present an IS hardware concept which is based on FPGA-technology for both sensor stimulation and data acquisition combined with adapted feature extraction methods for signal interpretation. Based on signal theory we are using a MLS signal, which we identified as an optimal signal for sensor stimulation and identification, respectively. This signal with a bandwidth up to $100\,MHz$ is generated by an FPGA and then reshaped and amplified in order to obtain a high quality MLS signal with variable amplitude. The sensor response is acquired by a high speed ADC in combination with the FPGA. Furthermore, the FPGA performs an FFT to obtain the impedance spectrum. Based on measurements with a commercial IS analyzer we transfer state-of-the-art feature extraction to sc-gas sensor applications. Relevant features are extracted by fitting impedance curves using equivalent circuit models. In order to develop a field test system that combines IS and temperature cycling of the sensor, we enhance and adapt the feature extraction algorithms. Information extracted from the curve shapes are significant features for gas discrimination.

This paper contains the results of two contributions to the International Workshop on Impedance Spectroscopy 2010, Chemnitz: "Low-cost impedance spectroscopy for semiconductor gas sensors—a hardware concept" and "Feature extraction based on impedance spectroscopy for semiconductor gas sensors".

Keywords: semiconductor gas sensor, FPGA based electronics, low cost, feature extraction

1 MOTIVATION

Due to their high sensitivity and low cost, semiconductor gas sensors are ideally suited for safety and security applications like fire and leakage detection (Conrad 2007), (Gramm 2003). The major drawbacks of this type of sensor are the poor selectivity and the sensor baseline drift. These major drawbacks must be overcome to achieve gas measurement system with reliable performance which is essential for safety and security applications. The main approaches to improve the selectivity and also the stability of sc gas sensors are temperature cycled operation (TCO) and electrical impedance spectroscopy (EIS) both combined with intelligent signal processing. In TCO mode the sensor temperature is changed periodically (A.P. Lee 1999), for example using sinusoidal modulation or, in our case, linear temperature ramps and plateaus. During this cycle the DC resistance of the sensor is read-out continually. For EIS the sensor temperature is usually kept constant and the sensor impedance is measured over a wide frequency range, e.g. during a frequency sweep (Sberveglieri 1995), (Schierbaum 1995), (U. Weimar 1995).

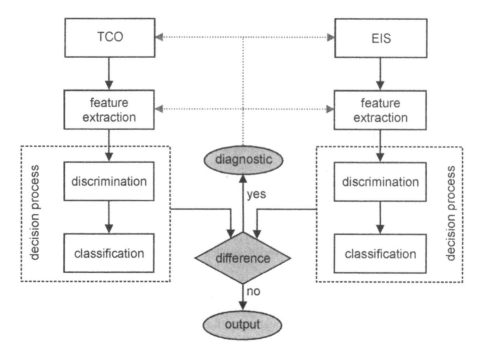

Figure 1. Self-monitoring and self-diagnosis strategy for applications with semiconductor gas sensors: Using a combination of TCO and EIS, redundant data from both methods are evaluated separately and the output is checked for differences.

Sensor poisoning of semiconducting metal oxide gas sensors is a serious problem in many safety and security applications. The poisoning, i.e. by irreversible gas adsorption blocking the surface, is usually a slow process. The sensor resistance degrades slowly over a period, while the sensor appears to be operating normally (K.F.E. Pratt 1997). However, using pattern analysis methods the detection of target gases or alarm situations is only possible with optimal system function, i.e. still matching the original calibration. The combination of TCO and EIS in one system allows acquisition of redundant information which can be used to detect and identify a system malfunction, i.e. caused by a damaged sensor, by separate signal evaluation and comparing results of both methods (P. Reimann 2008) (fig. 1). For the TCO adaptive and modular hardware was (T. Conrad 2006) developed in our group which is currently being used in various lab und field tests (P. Reimann 2009). EIS measurements on the other hand are currently only possible in a laboratory environment based on a commercial impedance analyzer and an optimized setup (T. Conrad 2007). Apart from the poor mobility of the impedance analyzer, this system is very expensive and also requires a long acquisition time for the sensor impedance. Therefore a combination of TCO and EIS is currently not possible. Especially for field tests a compact and low cost hardware has to be developed to make use of this approach. Closely connected with the hardware development and essential for complete measurementsystems is an efficient feature extraction strategy for the gas sensor data. This paper will present a concept and first results for both hardware and feature extraction, respectively.

2   FFT BASED MEASUREMENT CONCEPT

To improve and overcome the major drawback of the semiconductor (sc) gas sensors, i.e. the poor selectivity and stability, e.g. caused by sensor baseline drift, we are using TCO mode combined with intelligent signal processing. In TCO mode the sc sensor temperature is changed periodically over a certain time, the cycle duration. While the cycle duration is primarily determined by the sensor thermal time constants, depending on the sensor construction and assembly, the temperature range and cycle shape are chosen for optimal detection of the target gases for a given application. A special hardware was developed, which controls the sensor temperature and reads out the DC resistance during the temperature cycle. Besides the DC resistance sc gas sensors also show significant changes of their impedance when exposed to different gases and/or operating at different temperatures. Currently, the impedance is measured using lab equipment based on a commercial EIS analyzer (Solartron SI 1210). With this system an impedance sweep of 30 frequencies from $100kHz$ up to $32MHz$ is possible in approx. one minute. Due to the long acquisition time of the system, the sc sensor can only be operated in a static temperature mode. A combination of TCO and EIS cannot be realized with this setup.

To overcome this limitation and based on intensive studies (P. Reimann 2010) we developed a measurement concept allowing the promising combination of both methods with an EIS bandwidth up to $100MHz$. This overall concept is shown in fig. 2. The impedance of the device under test (DUT), i.e. the sc gas sensor, is determined by calculating the transfer function. Therefore the sensor is stimulated using a commercial signal generator with a broad frequency spectrum. The stimulation signal (U_{gen}) and the sensor response (U_{ref}) are measured using an oscilloscope. The impedance of the sensor is calculated by Fast Fourier Transformation (FFT) followed by dividing both spectra.

The major advantages of this approach are short measurement time and a relatively simple hardware setup. Even when using a very short signal (e.g. a rectangular pulse of some ns with a repetition rate of $100kHz$), the sensor is stimulated with all frequencies contained in the signal, i.e. with a broad signal bandwidth. The overall acquisition time is determined by the signal length, the acquisition time and the time for calculating the impedance spectrum, which is usually the longest.

After same tests with this setup and comparison of different stimulation signals (single pulse, white noise, Chirp, MLS) for EIS, we identified the Chirp and MLS signals as optimal

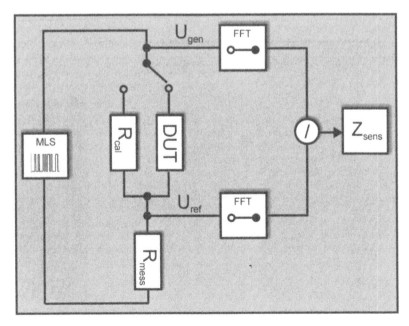

Figure 2. Overall measurement concept for EIS based on pulse stimulation and FFT transformation of the time domain response. The setup is calibrated by replacing the DUT with different calibration resistors.

for sensor stimulation (P. Reimann 2010). A Chirp is a sinusoidal signal with continuously increasing frequency; the MLS (maximum length sequence) is a type of pseudorandom binary sequence (J. Schoukens 2001). For efficient hardware implementation we decided to use the MLS signal, which requires only digital output instead of analog output for the Chirp signal.

The raw data provided by this measurement concept is the complex impedance spectrum, i.e. amplitude and phase of the sensor impedance over the signal bandwidth. A maximum length sequence (MLS) with a length of $2^m - 1$ can be generated by using a linear feedback shift register with a length of m. Mathematically, the generation of an MLS can be interpreted as coefficients of irreducible polynomials in a polynomial ring. For a sequence with maximum length the initial polynomial must be primitive. The resulting sequence contains exactly one high state more than the number of low states. Therefore the crest factor (peak to average ratio) is almost one. The spectrum of an MLS is scalable in frequency and has an envelope in shape of the sinc-function. The lowest frequency in the spectrum is adjustable with the length of the sequence, i.e. the value m; the highest frequency depends on the clock frequency of shift operation. Since tests with the described set-up based on signal generator and digital oscilloscope had shown promising results, we decided to develop a dedicated hardware based on this concept.

Field Programmable Gate Arrays (FPGAs) are very suitable for generation of MLS sequences. In a preliminary study we implemented an MLS-generator in a FPGA and analyzed the output signal in the time- and frequency domain. Figs. 3 and 4 show the time domain and FFT spectra of the theoretical, simulated and real MLS signals and their FFT spectra, respectively. The ideal signal was generated by MATLAB, the simulated signal is produced by simulating the signal-refining amplifier circuit with an equivalent LCR circuit in place of the gas sensor (Schierbaum 1995). The third signal is the measured FPGA output. The simulated and measured signals show a good correlation with the ideal signal. The signal is based on a clock frequency of $200 MHz$, following a minimum pulse length of $5ns$, and a sequence with the length of 4095.

Figure 4 shows the corresponding energy spectrum in the frequency domain. It is not possible to use the whole frequency range because the spectral density goes down and reaches

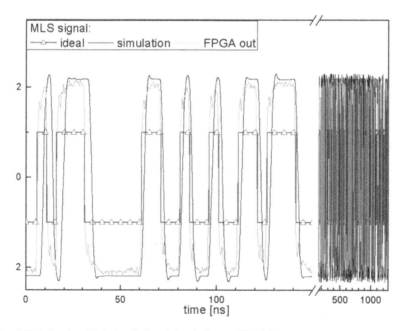

Figure 3.   MLS signals—ideal signal, signal simulation and FPGA output.

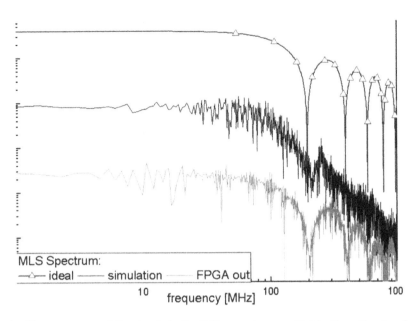

Figure 4. Energy spectra of the signals in fig. 3 (the spectra are offset in the y-direction for better clarity).

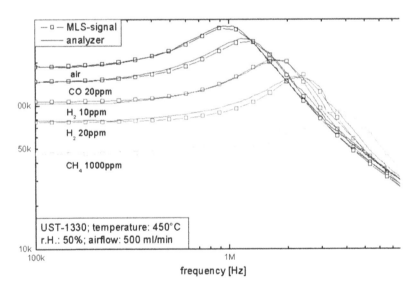

Figure 5. Comparison of EIS spectra obtained with the MLS-based approach and a commercial analyzer. Measurements were made with a commercial sc gas sensor (UST 1330) in humid air (50% R.H.) for pure air and four different gases as shown.

zero at $200 MHz$. A useable maximum frequency is $100 MHz$, at whichthe spectral density has decreased by $3dB$.

First measurements with a commercial sc gas sensor (UST 1330 (UST Umweltsensortechnik GmbH)) have shown promising results (fig. 5). The correlation between our FFT based concept and the results obtained with the commercial impedance analyzer is quite good with a maximumdeviation of 7%. Different gases cause different amplitude and phase response and also a shift of the resonance frequency. To separate these gases, significant features have to extract.

3 FEATURE EXTRACTION

In our tests two operating modes were used: the static temperature mode (STM) and the TCO mode. Whereas in STM the sensor temperature is constant duringmeasurement time, in TCM the temperature is changed and the impedance is measured. Both modes are measured with the FFT approach with a MLS signal length of $21\mu s$ and deliver significant impedance curves for feature extraction.

The features of EIS measurements are separated in primary and secondary features. The primary features are calculated from the raw data in Bode and/or Nyquist plots. Prominent examples for features extracted from both plots are slopes, mean values or characteristic measurement points like minima or maxima as shown in fig. 6. To extract the secondary features from measurement data an additional step is necessary in which these impedance curves are fitted for each sweep using equivalent circuit models (Gutierrez-Osuna 2002) composed of four parameters, i.e. two capacities, a resistor and an inductivity (fig. 11). The subsequent EIS signal evaluation is then based on the values of the extracted circuit elements. Both primary and secondary features were evaluated for gas discrimination using Linear Discriminant Analysis (LDA) (Gutierrez-Osuna 2002), (et al. 2001).

Measurement results were obtained using a commercial sc gas sensor (UST 1330), which was exposed to three different standard gases (CO, H_2 and CH_4), each with two concentrations (5, 10ppm for CO and H_2; 1000, 2000ppm for CH_4) in humid air (50% R.H.) supplied by an automatic gas mixing system (GMS). The three gasesused are typical target gases for applications in safety and security like fire detection (Conrad 2007). The sensor temperature is 325°C during STM measurements.

3.1 Static temperature mode (STM)

Discussion of the results is divided in primary and secondary features.

3.1.1 Primary features

Fig. 6 shows the Bode and Nyquistplot of the sensor raw data obtained with the FFT based concept. Different gases and gas concentrations cause different characteristics in both diagrams. For signal evaluation a comparison of these diagrams is necessary and yields different features. In fig. 6 the position of the maxima for real and imaginary parts, respectively, of the Nyquist plots are marked with yellow and orange markers for example. The corresponding values of both real and imaginary part are extracted as features and, in addition, the corresponding frequency and impedance values of these points from the Bode diagram.

A further set of primary features is extracted from the Bode diagram by dividing the frequency range into four sections and extracting the mean values and average slopes for each of these sections as shown in fig. 7.

3.1.2 Secondary features

Data obtained in STM have been fitted to an equivalent circuit (fig. 11) by using the software LEVM (Barsoukov 2005) and the values of R, C, C_p and L are used as features for the following discrimination.

To verify the fitted parameters, fig. 8 shows a comparison of the Nyquist diagram for the obtained equivalent circuit and the raw measurement data. The correlation is quite satisfactory proving both the appropriateness of the equivalent circuit model as well as the stability of the fitting algorithm.

3.1.3 Discrimination using LDA

Based on the extracted features a discrimination of the gases is carried out by using LDA, i.e. by calculating a projection into two dimensional space with optimum separation between groups and minimum scatter within groups. In our case, the groups represent the different gases and comprise the data for both gas concentrations. The LDA in fig. 9 is based on only the primary features and shows a good separation of the three measured gases and air

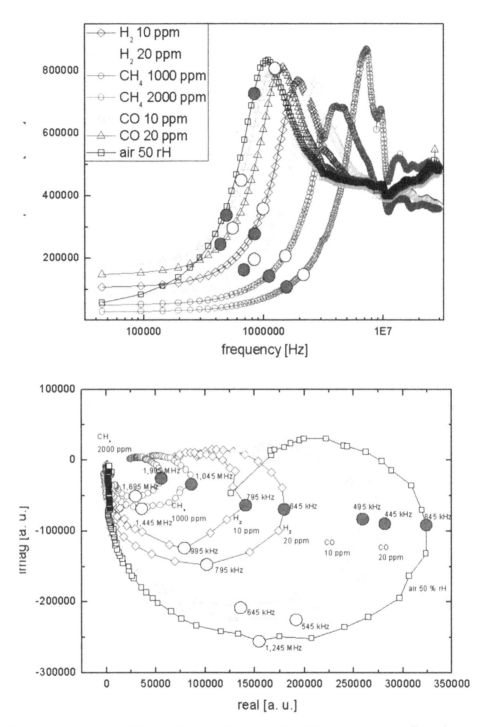

Figure 6.   Comparison of Bode and Nyquist diagrams with significant and corresponding points.

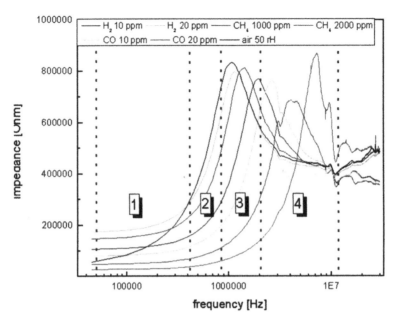

Figure 7. Primary feature extraction from the Bode impedance plot: the plot is divided into four sections from which the mean values and average slopes are extracted.

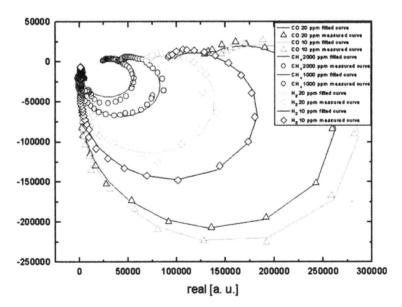

Figure 8. Comparison of the measured Nyquist diagram (symbols) and the Nyquist diagram calculated for the equivalent circuit based on the fitting parameters.

allowing identification of each gas independent of the concentration. In comparison, a much poorer gas discrimination is obtained when the four secondary features, i.e. the values of the equivalent circuits are used as shown in fig. 10. Here, separate groups are obtained for the two concentrations for each gas. In addition, the groups are less compact, probably reflecting the run in of the sensor after the start of the gas exposure until a steady state is reached. By

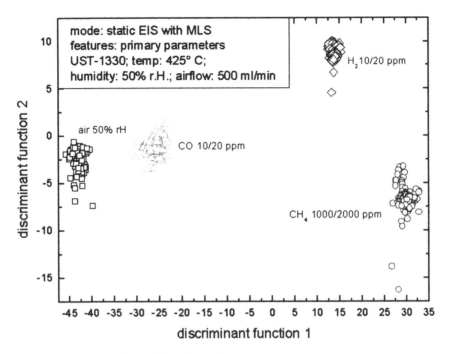

Figure 9. Separation by LDA based on primary features.

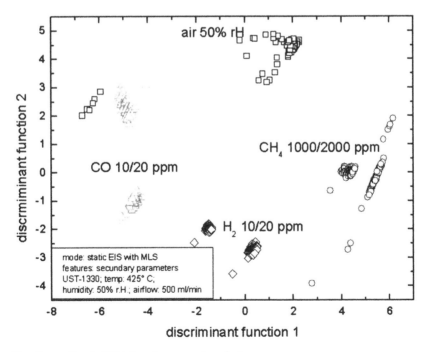

Figure 10. Separation by LDA based on secondary features.

combining both primary and secondary features a good quality LDA projection as obtained, comparable to the projection from the primary features only.

3.2 *T-cycling mode (TCM)*

For TCO the sc gas sensor is operated with a temperature cycle, which is optimized for underground fire detection (Conrad 2007) and is also suitable for our test gases. In the cycle with an overall duration of $70.5sec$ the sensor is operated between $185°C$ and $387°C$. During TCO a complete EI spectrum is obtained approx. every sec, resulting in 70 impedance spectra for each cycle. The spectra, however, have a lower frequency resolution to allow fast acquisition; two examples for pure air and $20ppm\ CO$ are shown in fig. 12 for a single temperature cycle each. The obtained Nyquist plots show significant differences between air and CO. The new challenge for the feature extraction of TCO data is find suitable features to describe or represent these patterns. Currently, the TCO mode only allows measurements with fewer sample points so that fitting to an equivalent circuit is not possible and the secondary features above mentioned cannot be extracted.

The best results for gas discrimination are obtained with descriptive features as shown in fig. 13, such as the area between the minimum and the maximum curve, the difference of the value at a defined frequency, the shift of the frequency maximum between certain curves or the averaging of the impedance over certain frequency ranges.

The LDA projection based on these features shows relatively poor separation between CO and H_2, with good discrimination vs. air and CH_4 as shown in fig. 14.

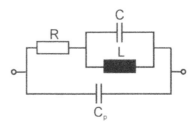

Figure 11.   Equivalent circuit model for fitting the measured impedance curves using the software LEVM.

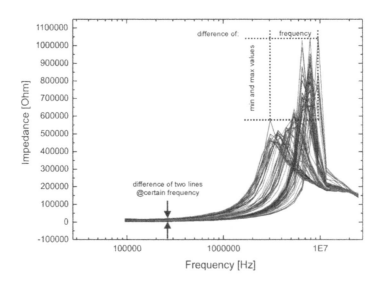

Figure 12.   Possible descriptive features to be extracted from impedance spectra obtained in TCO mode.

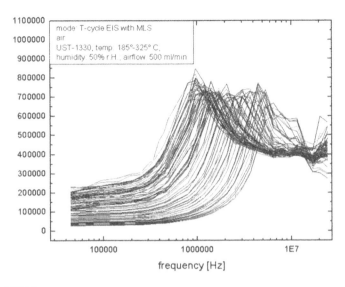

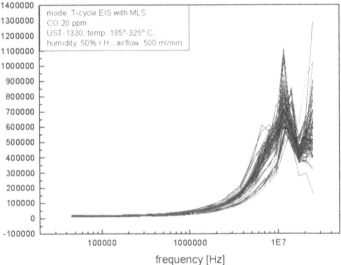

Figure 13.   Impedance spectra obtained in TCO mode for pure air (top) and 20*ppm CO* (bottom). Both plots show 70 complete spectra obtained during a single temperature cycle.

4   HARDWARE CONCEPT

The new low-cost hardware platform will be designed for a higher sample rate, so that both fitted and descriptive features can be obtained from the EIS spectra. Furthermore, this platform will combine the normal TCO mode with DC measurement and the new mode combining EIS and TCO to allow a self-monitoring of the system. Fig. 15 shows the overall hardware concept of the combined EIS/TCO system, which is based on an FPGA. Signal generation is performed by the FPGA combined with a separate circuit for signal shaping and selecting the MLS amplitude. Data acquisition in the time domain is achieved with a high speed ADC and subsequently transformed to the frequency domain by the FPGA, e.g. by implementing algorithms like FFT. The FPGA will also synchronize these processes to ensure data integrity, i.e. a fixed correlation between excitation signal and measured sensor response which is of fundamental importance for the following signal interpretation. On the left hand side of fig. 15, the circuit for signal refining/shaping and amplification is

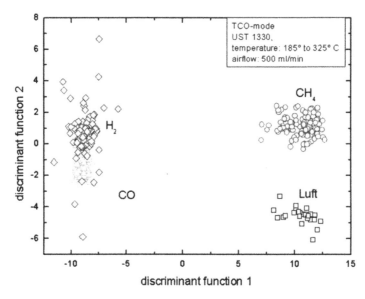

Figure 14. LDA projection for the 3 test gases (2 concentrations each) and air based on descriptive features obtained from EIS in TCO mode.

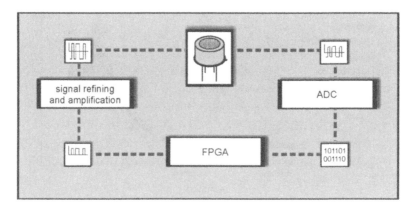

Figure 15. Overall hardware concept for a platform for high speed EIS data acquisition. This module will be integrated with a second platform controlling the TCO mode.

represented—this circuit basically consists of two RF differential amplifiers. The refining is done by the first amplifier, which is driven into saturation thereby transforming the imperfect low voltage differential signal (LVDS) output of the FPGA into a MLS signal with well-defined rectangular pulses. The second differential amplifier has a variable gain, providing an output of adjustable amplitude.

Synchronously, the high speed ADC captures the response of the gas sensor. The ADC is driven by a high performance, low jitter clock synchronizer in order to minimize phase noise and to enhance the quality of the analog to digital conversion. The FPGA provides an interface to import the data generated by the ADC and store them internally. After data acquisition, an FFT is performed using the radix-4 algorithm and the results are sent to the host PC for further processing. In the future, the signal analysis, i.e. LDA projection and classification of the data, should also be carried out by the FPGA, e.g. using a soft processor or by a separate micro-controller.

5 CONCLUSION AND OUTLOOK

In this paper we present a new hardware concept for low cost impedance measurements based on FPGA and MLS signals. The sensor impedance is calculated via sensor stimulation with MLS signals and obtaining the sensor response in the time domain. This approach has proven successful with standard lab equipment (signal generator for stimulation and oscilloscope for data acquisition). The acquired data are then transferred into the frequency domain where state-of-the-art feature extraction methods suitable for sc gas sensor data can be applied. We have shown that primary and secondary features extracted from static temperature mode allow good separation of different gases independent of concentration, e.g. using LDA. For the first time, this measurement approach also allows impedance measurements during temperature cycled operation (TCO). However, due to limited acquisition rate only primary features can currently be extracted which lead to sub-optimal gas discrimination.

In future we will optimize the hardware concept further and build up a system on a single PCB, which will combine TCO with DC and EIS measurement mode to allow a self-monitoring of the system. Furthermore, the implementation of other signal shapes for sensor stimulation, e.g. a Chirp signal, will also be studied. Closely connected with this hardware development, the signal processing including preprocessing, e.g. normalization or filtering, feature extraction, dimensionality reduction and classification will be developed to obtain highly selective and stable gas measurement systems.

REFERENCES

Barsoukov, E.; Macdonald, J. (2005). *Impedance Spectroscopy Theory, Experiment and Applications.* Wiley.

Conrad, T.; Reimann, P.S.A. (2007, Okt. 29–31). A hierarchical strategy for under-ground early fire detection based on a t-cycled semiconductor gas sensor. In *IEEE Sensors Conference*, Atlanta, USA.

Conrad, T.; Trmper, F.; H.H.A.S. (2007). Improving the performance of gas sensor systems by impedance spectroscopy: Application in under-ground early fire detection. In *Proc. SENSOR Conference*, pp. 169–174. AMA Service GmbH.

E.L. et al., (2001). Multicomponent gas mixture analysis using a single tin oxide sensor and dynamic pattern recognition. *IEEE Sensors Journal vol. 1 no. 3*, pp. 207–213.

Gramm, A.; Schtze, A. (2003). High performance solvent vapor identification with a two sensor array using temperature cycling and pattern classification. *Sensors and Actuators B 95*, pp. 58–65.

Gutierrez-Osuna, R. (2002). Pattern analysis for machine olfaction: a review. *IEEE Sensors Journal vol. 2, no. 3*, pp. 189–202.

Lee, A.P.; B.R. (1999). Temperature modulation in semiconductor gas sensing. *Sensors and Actuators B 60*, pp. 35–42.

Pratt, K.F.E.; D.E.W. (1997). Self diagnostic gas sensitive resistors in sour gas applications. *Sensors and Actuators B 45*, pp. 147–153.

Proc. Eurosensors (2006, Sep. 17–20). *A versatile platform for the efficient development of gas detection systems based on automatic device adaptation*, Volume XX, Gteborg, Sweden. Proc. Eurosensors.

Proc. IEEE Sensors Conf. (2009, October 25–28). *Field-test system for underground fire detection based on semiconductor gas sensors*, Christchurch, NZ. Proc. IEEE Sensors Conf.

Reimann, P.; Dausend, A.; A.S. (2008, Oct. 27–29). A self-monitoring and self-diagnosis strategy for semiconductor gas sensor systems. In *Proc. IEEE Sensors Conference*, Lecce, Italy.

Reimann, P.; Dausend, A.; A.S. (2010, 18.–19. Mai). Optimale signalanregungzur low-cost impedanzmessung von halbleitergassensoren. In V. Verlag (Ed.), *Sensoren und Messsysteme*, Nrnberg, pp. 18.–19.

Sberveglieri, G. (1995). Recent developments in semiconducting thin-film gas sensors. *Sensors and Actuators B 23*, pp. 103–109.

Schierbaum, W.G.K. (1995). Sno2 sensors: Current status and future prospects. *Sensors and Actuators B 26–27*, pp. 1–12.

Schoukens, J. (2001). System identification a frequency domain approach. *IEEE Press*.

Weimar, U.; W.G. (1995). A.c. measurements on tin oxide sensors to improve selectivities and sensitivities. *Sensors and Actuators B 26–27*, pp. 13–18.

Lecture Notes on Impedance Spectroscopy – Kanoun (ed)
© 2012 Taylor & Francis Group, London, ISBN 978-0-415-69838-2

Current source considerations for broadband bioimpedance spectroscopy

P. Annus & M. Min
ELIKO Competence Center, Tallinn, Estonia
Department of Electronics, Tallinn University of Technology, Tallinn, Estonia

M. Rist & J. Ojarand
ELIKO Competence Center, Tallinn, Estonia

R. Land
Department of Electronics, Tallinn University of Technology, Tallinn, Estonia

ABSTRACT: Classically bioimpedance measurements are performed by injecting current into the tissue under investigation, and raw data is acquired as measured response voltage. While it is relatively easy to design an optimized current source for single frequency or narrow bandwidth measurements, things are considerably more difficult when broadband spectroscopic measurements are required. Active current sources work reasonably well up to some megahertz, but unfortunately their parameters degrade significantly at higher frequencies. Working range of passive or purely resistive current sources on the other hand, while not comparable at lower frequencies can reach potentially higher frequencies. Imperfections caused by limited output impedance can be compensated by performing simultaneous current measurements. Comparison of three designs is given, results analyzed, and practical circuit tested.

Keywords: bioimpedance spectroscopy, current source, current measurement

1 INTRODUCTION

Excitation with current source for impedance measurement is generally preferable over sourcing voltage. Real current sources however have their limitations. It is important to distinguish between limitations in the current source circuitry itself, and adverse effects caused by parasitic components outside of the current source. Problems outside of the current source itself are mostly related to cabling and electrodes. Important component, omnipresent in all types of cabling is capacitance between two wires. In typical shielded wires (to minimize external disturbances) it can easily reach hundreds of picofarads, and by forming RC integrator together with high output impedance of the current source is very serious obstacle. Long known remedy is called active driving of the shield (Graeme 1973). Problem with active driving is that it can be very complicated when multielectrode system is considered, where each electrode can assume the role of both the excitation source and voltage pickup. Practical experiments show that instability is often very difficult to tackle with. Nevertheless in simpler systems actively driven shield, often together with another outer shield for noise reduction, is useful solution. In multielectrode environments usage of active electrodes is preferred, where current sourcing is placed into close vicinity of the electrodes. It should be noted that there is still residual parasitic capacitance in the order of some picofarads at the outputs. Partly because of PCB design, and used connectors, partly because some sort of protective clamping is usually needed. In current paper it is taken into account as 1pf capacitors from output to ground. Different set of problems and solutions are connected with the current source itself. From the modelling point of view real current source can be seen as Thévenin circuit, where ideal voltage source is connected series with high impedance, and as Norton circuit, where ideal current source is connected parallel to high impedance. Thévenin type of presentation leads to current source, which is probably easiest to construct. Good quality voltage sources can be made in wide range of frequencies and voltages, and by adding series resistor simplest current source is derived. Still several problems are apparent. Voltage source is limited by supply range. It can be doubled in bridged connection, but achieving voltages over some tens of volts is usually not feasible in embedded electronics. Resistor on the other hand needs to be usually much larger then impedance under examination, which could severely limit achievable current levels. Thermal noise contribution of resistors should be taken into account as well. With every resistor there is also inherent parallel parasitic capacitance, affecting output impedance at higher frequencies. In reality compromise can be achieved by using load dependent correction. It can be achieved by measuring real transfer characteristics with known loads, or alternatively by continuous measurement of the excitation current (Wang 2005). One strong benefit is related to safety. Since the resistor is connected in the worst case to the supply line, then strong upper limit exists for generated current. Also frequency dependent current limitation can be easily achieved (Annus 2005).

There are many different active current source designs. Usually they are based on operational amplifiers, and consequently appeared together with the first devices. One of the known designs from early sixties is the Howland current source. It was invented by Bradford Howland from MIT around 1962 (Pease 2008). Historically it has been described as very clever circuit, which is almost useless (Horowitz 1989). Nevertheless several modifications of the Howland circuit have been reported to be used successfully for impedance measurement, like (Chen 2007), (Hong and Bayford 2007), (Bertemes-Filho 2000). Second common type is the so called load in the loop circuit, whereas impedance under measurement is essentially placed in the feedback loop of an inverting amplifier, and the current in the load is proportional with the input voltage of the amplifier (Boone 1996), (Annus 2008), etc. Third design is centred on current mirroring, and is mostly used in chip design (Kasemaa 2008), (Min 2006). In fact there are many other possibilities, such as supply current sensing based current sources, different multiamplifier designs, circuits based on so called diamond structures (Min 2006), current conveyors, already mentioned current measuring and correction circuits (Wang 2005), and others. General discussions on wide bandwidth current source parameters can be found in (Seoane 2006), and discussion of the load in the loop circuit in (Annus 2008).

Current paper considers and compares three designs. First revisits well-known design based on transor (Min 2006) OPA660 from Texas Instruments using newer derivative OPA860. Second is simple active current source with Analog Devices AD8130 differential-to-single-ended amplifier, based on (Birkett 2005) and third is simple resistive current source.

2 CURRENT SOURCES

Transor based current source seems to be an attractive choice. Unfortunately older chips had rather limited output impedance. New OPA860 has improved parameters. Therefore it seems reasonable to test an active current source according to Fig. 1.

Probably one of the best active current source designs is based on AD8130 amplifier from Analog Devices, and is included for comparison on Fig. 2.

Passive, resistive current source is considered on Fig. 3. It consists of one resistor R_S, together with inherent parasitic capacitances C_S, and C_P.

Following figures introduce simulated current magnitude and phase characteristics of examined current sources at different load impedances.

As it can be seen on the Fig. 3, and also on the magnitude and phase characteristics of the passive resistive current source (Fig. 8, 9) parallel parasitic capacitance C_S will lower the output impedance at higher frequencies, and consequently rise output current. It is relatively easy to correct by introducing two more components, resistor R_{S1}, with its parallel parasitic capacitance C_{S1}, and compensating capacitor C_K Fig. 10.

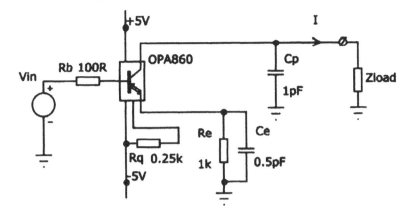

Figure 1. Simplified active current source with OPA860 amplifier.

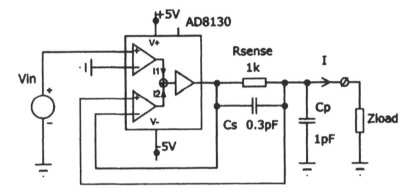

Figure 2. Simplified schematic of an active current source with AD8130 amplifier.

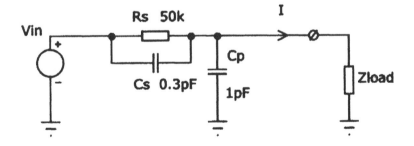

Figure 3. Passive, resistive current source.

Magnitude and phase characteristics of this modified current source can be seen on Fig. 11, and Fig. 12.

3 NOISE CONSIDERATIONS

Current noise of these three circuits is compared on following Fig. 13, and Fig. 14. It is clear that simple resistive current source has best noise parameters of all these three designs. Worst of all is OPA860 based current source.

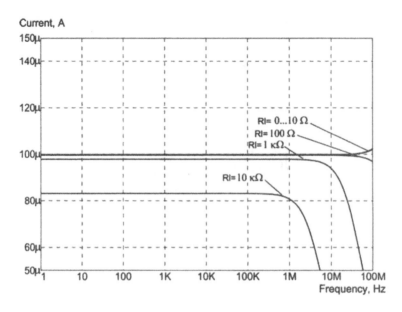

Figure 4. Magnitude characteristics of the output current of the transor based current source, at different load impedances.

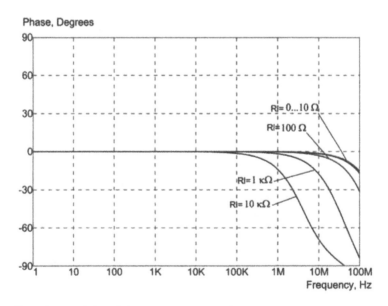

Figure 5. Phase characteristics of the output current of the transor based current source, at different load impedances.

4 PRACTICAL MEASUREMENTS

In addition to simulation resistive current source has been also measured with Wayne Kerr impedance analyser 6500B (Fig. 15). Both leaded resistors were measured, and surface mount device (SMD) resistors on printed circuit board (PCB) (Fig. 16). For comparison an impedance of an empty PCB was also measured, resulting parasitic capacitance calculated by best fitting was in the range of $0.07pF$. Typical parallel parasitic capacitance of the resistor was

66

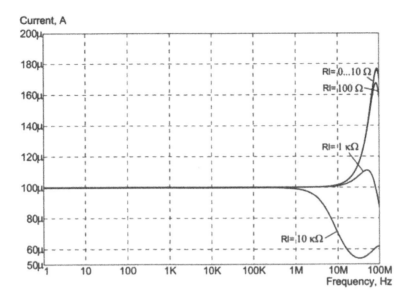

Figure 6. Magnitude characteristics of the output current of an active current source with AD8130 amplifier, at different load impedances.

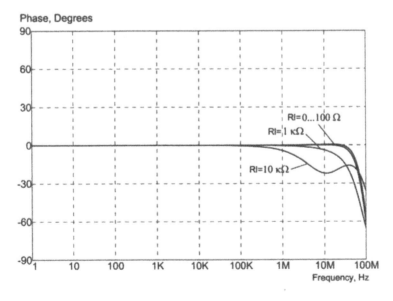

Figure 7. Phase characteristics of the output current of an active current source with AD8130 amplifier, at different load impedances.

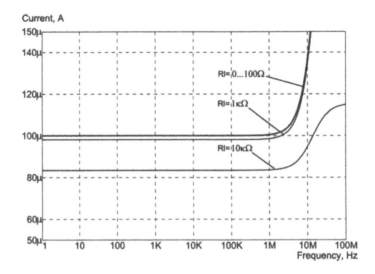

Figure 8. Magnitude characteristics of the output current of simple resistive current source, at different load impedances.

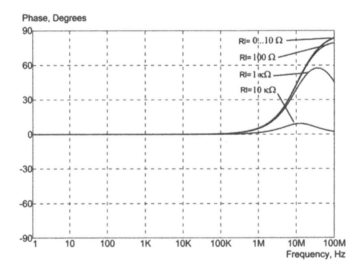

Figure 9. Phase characteristics of the output current of simple resistive current source, at different load impedances.

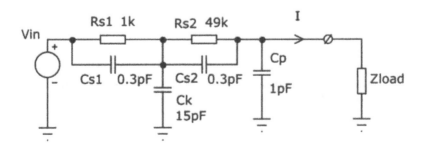

Figure 10. Modified simple resistive current source.

68

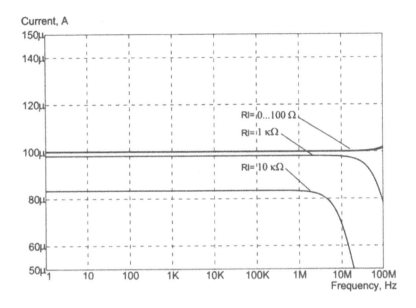

Figure 11. Magnitude characteristics of the output current of the modified simple resistive current source, at different load impedances.

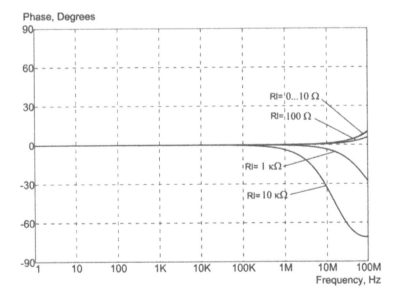

Figure 12. Phase characteristics of the output current of the modified simple resistive current source, at different load impedances.

69

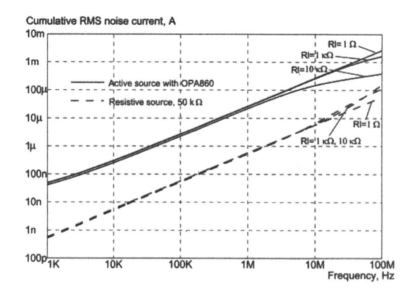

Figure 13. Comparison of output current noise of transor (OPA860) based current source and simple resistive current source, at different load impedances.

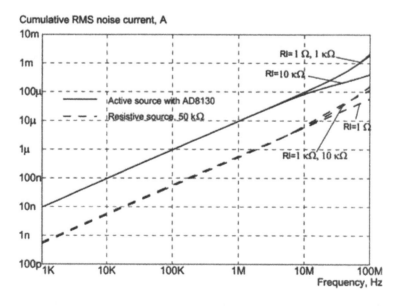

Figure 14. Comparison of output current noise of active AD8130 based current source and simple resistive current source, at different load impedances.

Figure 15. Measurement of the actual impedance of the leaded resistor, and SMD resistors mounted on FR4 board.

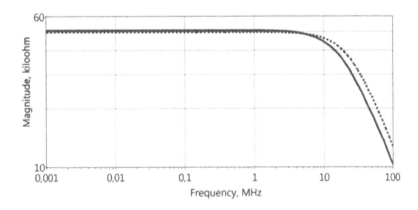

Figure 16. Measured actual impedance curve of the $49.9k\Omega$ leaded resistor, dotted curve, and $51k\Omega$ SMD resistor on FR4 PCB.

0.14–$0.16pF$. Series connection of two SMD resistors on PCB did not improve the results considerably. Slight preference should be given to leaded resistor

5 DISCUSSION

Frequency dependence of the output impedance of the current source is probably the most important parameter. For comparison simulated output impedance of all discussed current sources is drawn on common Fig. 17.

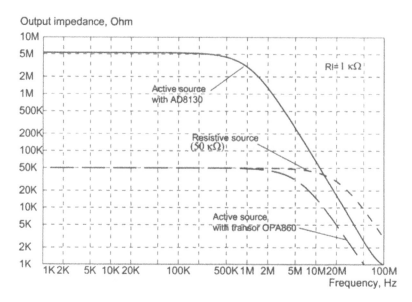

Figure 17. Comparison of the output impedances of discussed current sources.

From Fig. 17 it is evident that transor based current source is still not very viable solution. If frequency range from kilohertz to some tens of megahertz together with load impedances from some ohm to kilo ohm is considered, then simple resistor is a very strong contender. While having much lower impedance at low frequencies, than active current source, $50k\Omega$ versus $5M\Omega$, it behaves better at higher frequencies, can be introduced very near to the electrodes, and errors due to limited impedance are still negligible and can be easily compensated in software. It is less noisy, consumes much less energy and is considerably more stable and simpler then any active current source can be made. Slight modification can even further improve its characteristics at high frequency if required. In fact actual measurements gave slightly better results than anticipated during modelling.

ACKNOWLEDGMENT

This research was supported by the European Union through the European Regional Development Fund, and also by Enterprise Estonia through the Competence Center ELIKO.

REFERENCES

Annus, P., K.A.M.M.P.-T. (2008). Current source for bioimpedance measurement applications: Analysis and design. *Instrumentation and Measurement Technology Conference Proceedings*, pp. 848–853.

Annus, P., L.J.M.M.P.-T. (2005). Design of a bioimpedance measurement system using direct carrier compensation. *Proceedings of the 2005 European Conference on Circuit Theory and Design, ECCTD III*, pp. 23–26.

Bertemes-Filho, P., B.B.H.W.-A.J. (2000). A comparison of modified howland circuits as current generators with current mirror type circuits. *Physiol. Meas.*, pp. 1–6.

Birkett, A. (2005). Bipolar current source maintains high output impedance at high frequencies. *EDN*, pp. 128–130.

Boone, K.G., H.D.S. (1996). Approaches to analogue instrumentation design in electrical impedance tomography. *Physiol. Meas.*, pp. 229–247.

Chen, C.-Y., L.Y.-Y.H.W.-L.C.K.-S. (2007). The simulation of current generator design for multi-frequency electrical impedance tomograph. *Proceedings of the 28th IEEE EMBS Annual International Conference*, pp. 6072–6075.

Graeme, J.G. (1973). *Applications of operational amplifiers: third-generation techniques.* McGraw-Hill.

Hong, H., R.M.D.-A. and Bayford, R.H. (2007). Floating voltage-controlled current sources for electrical impedance tomography. *Circuit Theory and Design 18th European Conference*, pp. 208–211.

Horowitz, P, H.W. (1989). *The Art of Electronics (Vol. I)*. Cambridge University Press.

Kasemaa, A., A.P. (2008). Cmos current source for shortened square wave waveforms. *Electronics Conference*, pp. 848–853.

Min, Parve, M., T.S.-R. (2006). *Biomedical Electronics: in Wiley Encyclopedia of Biomedical Engineering.* John Wiley & Sons Inc.

Pease, R.A. (2008). A comprehensive study of the howland current pump. *National Semiconductor Corporation*.

Seoane, F., B.R.L.K. (2006). Source for multifrequency broadband electrical bioimpedance spectroscopy systems. a novel approach. *Engineering in Medicine and Biology Society 28th Annual International Conference of the IEEE*, pp. 5121–5125.

Wang, C., X.M.W.H. (2005). Mixing frequency biology impedance measurement with virtual reference point. *Instrumentation and Measurement Technology Conference, IMTC, Proceedings of the IEEE*, pp. 1407–1410.

Lecture Notes on Impedance Spectroscopy – Kanoun (ed)
© 2012 Taylor & Francis Group, London, ISBN 978-0-415-69838-2

Handheld impedance measurement platform for preliminary analysis in biomedicine and technology

Marek Rist
ELIKO TAK Ltd., Tallinn, Estonia

Paul Annus
Department of Electronics, Tallinn University of Technology, Tallinn, Estonia

Tonu Jaansoo
Innotal Ltd., Tallinn, Estonia

ABSTRACT: Typical impedance measurement process in scientific studies involves the use of dedicated laboratory equipment like Agilent 2494A or Wayne Kerr 6500B. Although precise and reliable, these kinds of instruments limit the measurements into dedicated workplaces due to their size, weight and cost. While it is relatively easy to design impedance measurement application for a specific task, it still often requires too much time and resources. The aim of this work is to provide a flexible and cost effective solution for performing impedance measurements both inside the laboratory and out. The design is targeting mostly the fields of material study, study of electrochemical cells (like batteries) and bio-impedance.

Keywords: impedance spectroscopy, expandable functionality, frequency stepping, multi-sine excitation, chirp excitation, current source

1 INTRODUCTION

Often there has surfaced a need to perform impedance measurements of various objects in an environment, where the use of laboratory grade equipment is not feasible due their size, weight and cost.

Development of application-specific hardware is expensive and time-consuming. This may lead to abandoning of new ideas in using impedance spectroscopy.

A search was made to find a handheld device for electrical impedance measurements. The ones found were highly application specific (battery diagnosis, ICG or LCR meters). None of those could be used for extracting actual impedance information in interesting frequency ranges. There have been a plenty of papers written on portable impedance analysers (Yang 2007). Unfortunately none of them has become more than a journal article. In order to perform impedance measurements, a real device was a necessity.

This article will concentrate mostly on the design of the measurement channel. User interface, power supply, communications and other similar sections will be excluded.

2 REQUIREMENTS AND SYSTEM SETUP

The major impedance related studies in TUT Institute of Electronics are related to biological materials and batteries. Other requests for instrumentation have included material study, study of plants for agriculture, monitoring for food processing etc.

Biomedical research usually focuses in the frequencies between $100Hz$ and $10MHz$ while the Z being in the range of 100Ω and $1k\Omega$. On the other hand, the interesting frequency range while measuring batteries is between $1mHz$ and $10kHz$, while the magnitude of impedance remains usually under 1Ω (H.-R. Trnkler 2007). Cell potential with the combination of low impedance requires potentiostatic operation.

Measurement of temperature of the "object under test" is critical because impedance is more-or less temperature dependant. User interface and primary processing must be implemented to offer stand-alone operation and additional Bluetooth/USB interface and command-set for controlling the device through PC.

A capability to save measurement data will also be implemented.

Proposed solution for the key elements of the system is shown in Fig. 1.

2.1 *Excitation waveforms*

One of the most important questions in impedance measurement is the type of excitation signal to use. The most accurate results are achieved by measuring with single frequency sinusoids. When the nature of the object is known, and measurement time is important, multisine measurements can be considered one of the best options. In the case where wide spectrum must be measured in short time, chirping should be considered for the task (Mart Min 2010).

In Biomedicine, good crest factor is necessary since the excitation applied to the tissue is highly regulated and this means that the energy must be concentrated on the frequencies that are being measured to achieve the best signal to noise ratio. To achieve good crest factor, rectangular forms of described signals can be used (Mart Min 2010).

The generation of the pre-mentioned signals will be performed using a sub-section shown in Fig. 2.

2.2 *Excitation coupling*

Various coupling methods (voltage divider (VD), voltage controlled current source (VCCS), auto balancing bridge etc.) are being used in impedance measurements. The main difference being: is the excitation realized by applying voltage across the sample and measuring current

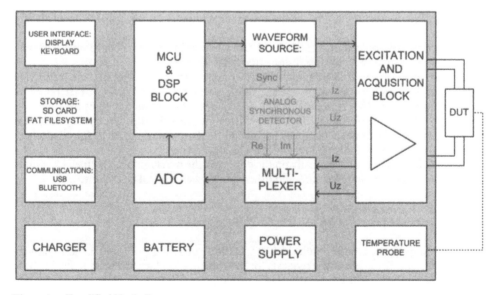

Figure 1. Simplified block diagram.

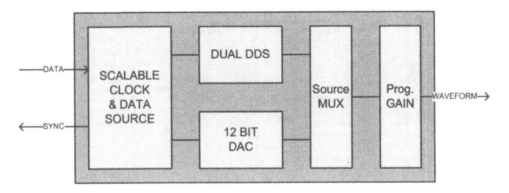

Figure 2. Waveform source.

induced in the sample, or applying current and measuring voltage drop across the sample (Technologies 2009).

Both methods have positive and negative sides.

Since major target-areas of the device are the measurement of bio-impedance and batteries, it was decided, that the best approach was to use a current source for excitation. This is because current source has high output impedance and it does not disturb the potentials across the object that is being measured. For an example, when impedance cardiograph is performed on a patient with impedance analyser that has voltage excitation, then electrical cardiograph signals are short-circuited by the low output impedance of the voltage source used in impedance measurements. Additionally, use of the current source offers a protection against over-currents.

Simplest and often best current source is resistor (Fig. 3A). Unfortunately, since we are dealing with portable battery operated equipment with limited voltage supply and we need to measure with currents up to $100mA$, this cannot be realized properly.

With active current sources, main topologies are operational trans-conductance amplifiers (OTA, Fig. 3B) and current sources based on operational amplifiers (OPA) utilizing negative feedback to control the current (U. Tietze 2008).

OTA's are fairly stable when driving reactive loads and have a wide bandwidth. At the moment, best available OTA can be considered to be OPA861. OPA861 features $\approx 80MHz$ bandwidth and $\approx 54k\Omega$ output resistance. Unfortunately the maximum output current $\pm 10mA$ and collector output compliance of $\pm 4V$ makes it not suitable for this task.

Standard OPA's can be used as VCCS by converting the output current to voltage by using known reference resistor and using it for the negative feedback to the amplifier (Fig. 3C, D, E). The current source in Fig. 3D illustrates this topology well. Unfortunately the use of instrumentation amplifier within the feedback loop lowers the usable bandwidth of the configuration shown in Fig. 3D when compared to VCCS in Fig. 3C and Fig. 3E.

Current source topology shown in Fig. 1C is fairly stable and usable up to MHz range. Downside of Fig. 3C is that the load will be floating and good common mode rejection ratio would be required from the acquisition amplifier.

Current source shown in Fig. 3E, with external output current booster, was the one used for this project due to very good bandwidth and non-floating load configuration.

The output impedance of the current source can be seen in Fig. 4 and also the realistic output impedance after implementation as a result of parasitics. These charts state, that high output impedance is impossible to achieve at high frequencies. Part of the output current will be shunted from the device under test and the measurement result will be erroneous. The difference in output impedances with 50Ω and 500Ω shunt resistors is negligible from $1kHz$ range and higher. This means, that the use of different shunt resistors to increase the output impedance of the current source can be considered only at the spectrum determined by the parasitic output capacitance.

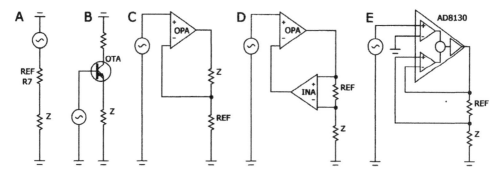

Figure 3. Various current sources.

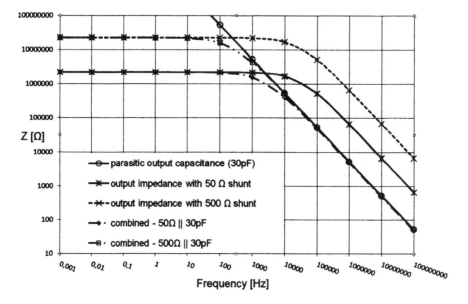

Figure 4. Output impedance and effect of parasitic output capacitance ($C = 30pF$).

For that reason, only one shunt was used for the current source.

To measure the actual current that flows through the sample, a trans-impedance amplifier was added to the return path of the excitation current. This does not improve the characteristics of the current source, but it enables to calculate the impedance of the DUT precisely, when sufficient ADC resolution is maintained. The achieved 1% accuracy impedance measurement range is increased (Fig. 5).

Both Li-Ion batteries and bio-impedance measurements can now be performed by using current source for excitation.

To be able to measure also high impedance range, the current source was made automatically reconfigurable to voltage source. Since voltage source has low output impedance, the parasitic capacitances at the output do not affect the bandwidth drastically. This option allows to measure impedances up to $10M\Omega$.

The combined topology designed to couple to the measured object is shown in Fig. 6.

2.3 Acquisition

Since the frequency range of the measurement device is being pushed near $10MHz$, the bandwidth of the acquisition amplifier must be at least $10MHz$ and preferably even

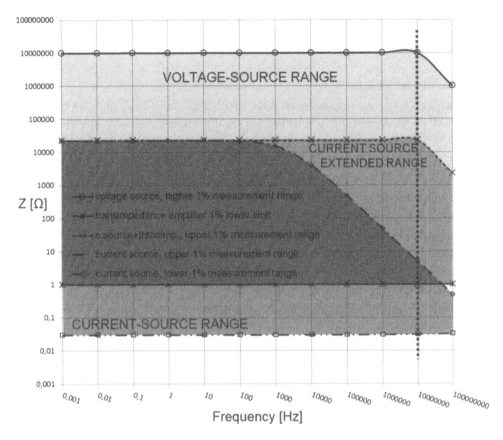

Figure 5. Measurement ranges.

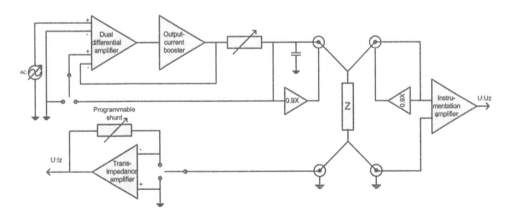

Figure 6. Coupling to the DUT.

$50MHz-100MHz$. The voltage drop across the DUT must be measured without disturbing the DUT and excitation source. This means that the acquisition amplifier must have high input impedance and wide bandwidth. Differential amplifiers have wide bandwidth but low input impedance, instrumentation amplifiers on the other hand have high input impedance but low bandwidth (U. Tietze 2008). There are few available instrumentation amplifiers that can be used near 1*MHz* range like AD8253, AD8220 and INA128 but this is inadequate for this design. A better instrumentation amplifier can be constructed from separate high performance operation amplifiers (Fig. 7).

2.4 *Detection and analysis*

The measured signals are sampled with 18bit ADC to be able to measure both 3,6V DC and few *mV* of AC in case of Li-ion battery. This is the simplest approach since the need to measure in *mHz* region prevents the use of analog filters for removing the DC component (Fig. 8).

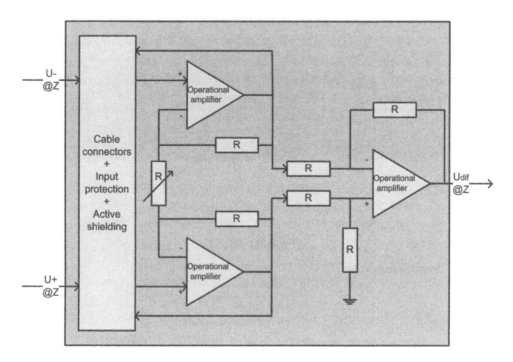

Figure 7.   Acquisition amplifier.

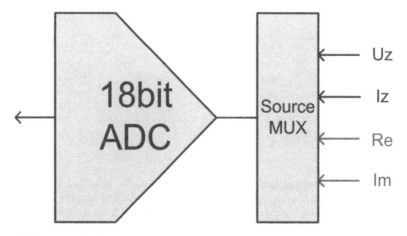

Figure 8.   Analog to digital conversion.

The method for detecting real and imaginary parts is DFT in case of sinusoidal sweep and multisine and FFT when excitating with chirp signals. There is also an analog synchronous detector implemented by using fast FET switches (E. Brigham 1988).

As the analog measurement channel contributes to frequency dependent magnitude and phase errors, the calibration plays an important role in the system. During the calibration, phase of the measurement channel is measured using a known reference. Phase calibration is realized during the detection by offsetting the phases of the synchronisation signals. The magnitude is being compensated after extraction of the real and imaginary components.

The device features a temperature probe to save temperature data alongside the impedance of the object.

Measurement data can be saved to the memory card or uploaded to a PC or a smartphone with USB or Bluetooth connection. User can also control the functions of the device through serial connection with Labview or custom software.

3 CONCLUSIONS

A handheld impedance measurement platform with frequency range from $10mHz$ up to $1MHz$ with (<1% error) extending to $10MHz$ (1...10% error) and with measureable impedance magnitude between $\approx 10m\Omega$ up to $10M\Omega$ has been designed for manufacturing.

Different excitation signals and configurability makes this a valuable tool in the field of impedance spectroscopy.

The device is being further improved with the optimization of measurement electronics and measurement algorithms.

ACKNOWLEDGMENTS

This research was supported by the European Union through the European Regional Development Fund and by Enterprise Estonia through the Competence Center ELIKO.

REFERENCES

Brigham, E. (1988). Fast fourier transform and its applications. *Prentice-Hal*.

Mart Min, e. a. (2010). Broadband spectroscopy of a dynamic impedance. *IOP Journal of Physics: Conference Series 224*, 1–4.

Technologies, A. (2009). *Impedance Measurement Handbook*.

Tietze, U., C.S. (2008). *Electronic Circuits: Handbook for Design and Application*. Heidelberg: Springer.

Trnkler, H.-R., Kanoun, O., M.M.M.R. (2007). Smart sensor systems using impedance. *Estonian Journal of Engineering 13*, 455–478.

Yang, e. a. (2007). Design and preliminary evaluation of a portable device for the measurement of bioimpedance spectroscopy. *Physiological Measurement 27*, 1293–1310.

Lecture Notes on Impedance Spectroscopy – Kanoun (ed)
© 2012 Taylor & Francis Group, London, ISBN 978-0-415-69838-2

Design and application of low cost system based on planar electromagnetic sensors and impedance spectroscopy for monitoring of nitrate contamination in natural water sources

M.A. Md Yunus
School of Engineering and Advanced Technology, Massey University, Palmerston North, New Zealand

S.C. Mukhopadhyay
Control and Instrumentation Engineering Department, Universiti Teknologi Malaysia, Skudai, Malaysia

ABSTRACT: A low cost system that measures impedance (real part and imaginary part) was used along with a planar electromagnetic sensor to estimate the amount of nitrate contamination in natural water sources. The system can generate alternating $10Volt$ peak-to-peak sine waveform signals at frequencies of $164kHz$, $260kHz$, $350kHz$ and $800kHz$. The design and operating principle of the low cost system as well as the different design of planar electromagnetic sensors have been reported in this paper. Every sensor was tested with nitrates forms namely, sodium nitrates ($NaNO_3$) and ammonium nitrates (NH_4NO_3), each of different concentration between $5mg$ and $20mg$ dissolved in 1$litre$ of distilled water and the real part and imaginary part of the sensors impedance were observed and analyzed using electrochemical impedances spectroscopy approach. For every aqueous solution type (sodium nitrates ($NaNO_3$) or ammonium nitrates (NH_4NO_3)) a semi-empirical equation based on multilinear regression model was determined from the percentage changes of real part and imaginary part values as compared to the mean real part, R_{total} and imaginary part, X_{total} calculated from mili-q water at the frequencies $260kHz$ and $800kHz$ and estimated concentration (EC) values. The semi-empirical equation could be used to determine the predicted concentration (PC) values of nitrates contamination (sodium nitrate or ammonium nitrate form) in natural water source. From the results and MLR models of the sensors, the best sensor was determined and selected to be used in detecting nitrate contamination of nitrate in the form of sodium nitrate and ammonium nitrate individually. The system and approach presented in this paper has the potential to be used as a useful tool for water sources monitoring in water sources.

Keywords: water quality monitoring, nitrates, planar electromagnetic sensors, low cost system, phase and gain measurement, AD8302, waveform generator, XR2206, electrochemical impedance spectroscopy, multilinear regression model

1 INTRODUCTION

The applications of impedance spectroscopy concept in corrosion measurement in material properties, chemical processes characterization, bioprocessing and medical diagnostic are being introduced at an increasing rate. The applications rely on the principle of applying time varying electrical signal applied upon a material or process of interest such as soil or chemical reaction to measure the resulting impedance and then interpreting the results into information such as material properties or the understanding of the reaction mechanism.

Looking at the versatility, simplicity, strength and advantages of impedance spectroscopy applications, we endeavoured to develop a low cost system in observing the reaction of planar electromagnetic sensors with water samples where the measurement is carried out in the frequency domain using the low cost system to estimate the contamination in natural water sources, in particular nitrate. The next section will discuss the motivation of developing a low cost system nitrate detection using planar electromagnetic sensors and analysis based on electrochemical impedance spectroscopy approach. The concept of multilinear regression model to estimate nitrate contamination using planar electromagnetic sensors is also discussed in this report.

1.1 *Importance of nitrate detection in drinking water sources*

Water supplies come from two main sources of surface and ground water. In general, the harmful substances that polluting the water supply may possibly come from industrial chemical production and metal-plating operations, pesticide or fertilizer runoff from agricultural land, and animal feedlot wastes (Stanley 2009). This research is particularly focusing on the sensing of Nitrates substances for natural water sources.

Nitrates are soluble chemical and can move with peculating and runoff water into open water such as big rivers and lakes, in this situation, the nitrates concentration will be disperse and posed no such effect or threat to the environment or living creatures, respectively. However, shallow pond that watered by the runoff from the heavily fertilized agricultural land or urine patch of animal feedlot (Di and Cameron 2007), may contain heavy amount of nitrates. Stanley in (Stanley 2009), reported that the wastes from animals feedlot are the major pollution of nitrates. It is the amino nitrogen present in nitrogen-containing wastes product such as cattle urine. In most cases, deep pond or well is usually nitrate free. However, in the long run, the amount of Nitrate may accumulate and problems such as Nitrate leaching into sources of drinking water may turn into negative impacts to the environment, livestock and human health. Nitrate (NO_3^-) is contaminant and may turn toxic to livestock including infants. This is explained when livestock especially ruminants and infants drink contaminated water with high level of Nitrate, the chances of Nitrate to turn into Nitrite (NO_2^-) is high due to both having high concentration of nitrate metabolizing triglycerides in their system (N.F. Metcalf 1987).

This condition when untreated or avoided, make them vulnerable to Methemoglobinemia, a disorder characterized by the reduction of blood capacity in carrying oxygen to a dangerous level and can cause organ tissues damage. The level of Nitrate consume by human may not exceeded $10mg$ $(NO_3^-) - N/L$ (Terblanche 1991) and for most farms, drinking water is around 2 to $3mg$ $(NO_3^-) - N/L$ and should not exceed $100mg$ $(NO_3^-) - N/L$.

Applications in the area of nitrate detection using indirect methods are associated with good sensitivity, excellent response and accuracy. Such methods are the detection of ammonium ion resulting from the reduction of nitrate by trichloride in hydrochloric acid (S.-J. Cho 2002), Spectrophotometric (Ferree and Shannon 2001), and Polarography and Voltmetric (Gumede 2008). However, these methods are commonly known to be expensive and involve tedious measurement steps which consume a lot of time, require controlled working condition, and preparation of extra reagent or chemical.

Another group of technology is linked with unsophisticated sensor fabrication process, simple monitoring instrumentation, fast and rapid response, accurate, compact and acceptably responsive even to limited amount of sample, and suitable for continuous measurement. They are commonly known as direct methods. Such applications are chromatography (Connolly and Paull 2001) and biosensors incorporating enzymes, antibodies, and whole cells (M.A. Md Yunus 2010). The most popular is potentiometry based on ion selective electrodes (ISE) (T. Kjr 1999)/(Bendikov and Harmon 2005). However, the strength of the output signal depends on the strength or concentration of the targeted ions and often requires amplification. Moreover, the output signal also susceptible to the interference from not targeted ions thus a large number of reagents were necessary for the purpose of neutralization of the noise and adversely might be harmful to the environment (S.-J. Cho 2002)/(M.J. Moorcroft 2001). To date, there are very few applications related to the use of electrochemical impedance spectroscopy approach in nitrate detection in natural water sources. Therefore, this research motivates to develop a low cost system based on planar electromagnetic sensors, analyze the results using impedance spectroscopy method and a multilinear regression was used for the empirical equation development.

2 PLANAR ELECTROMAGNETIC SENSORS

In this research, the planar electromagnetic sensors were designed using Altium Designer 6. The sensor was fabricated using simple printed circuit board (PCB) fabrication technology (thickness of 0.25*mm*). The sensor is then coated using Wattyl Incralac Killrust aerosol to form a coating layer in order to prevent any direct contact with water sample. Next section will discuss the construction and operating principle of the planar electromagnetic sensors.

2.1 *Construction and operating principle of sensors*

Previous applications have demonstrated that planar electromagnetic sensors are used for non-destructive testing of conducting and magnetic type materials (Goldfine 1993)/ (N.J. Goldfine and Lvett 1995)/(Mukhopadhyay 2002) and it is shown that the properties of dielectric materials have a great influence on the output response of the sensor (Mukhopadhyay 2002). A method for inspecting integrity of different coins and successfully discerns between coins is introduced in (D. Karunanayaka and G. 2006). The application of the new planar electromagnetic sensors for nitrate detection in natural water sources using impedance spectroscopy approach is considered in this report.

Figure 1 depict the top view and bottom view of series combination sensors: SCS1 (case (a) of figure 1), SCS2 (case (a) of figure 1), SCS3 (case (a) of figure 1) and SCS4 (case (a) of figure 1). Each one of all the sensors consists of a meander coil spiraling inwards in rectangular form and the inner end is connected to an interdigital sensor completing a serial connection. Table 1 summarizes the description of the sensors. The meander type of coil is connected in series with the interdigital coil and an ac voltage is applied across the combination of the coils. The meander coil produces a magnetic field and the interdigital coil produces an electric field. The combination of meander and interdigital coils produce electromagnetic field which interacts with the material under test. The purpose of providing a grounded backplane is to

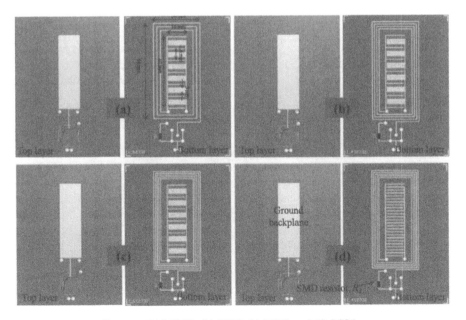

Figure 1.   Schematic diagram of (a) SCS1, (b) SCS2, (c) SCS3, and (d) SCS4.

Table 1. Parameters of planar electromagnetic sensors combined in series.

Location in coordinates sensor type	Parameters			
	Length (*mm*)	Width (*mm*)	Number of turn (*N*)	Width of negative electrode (*mm*)
SCS1	54	26	3	3
SCS2	54	26	4	3
SCS3	54	26	5	3
SCS4	Similar with SCS3 except the sensor has a positive electrode followed by a series of three negative electrodes. The above pattern was repeated with a driving electrode again.			

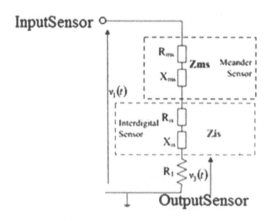

Figure 2. Electrical equivalent circuit of the planar electromagnetic sensor.

Figure 3. Experimental setup for the experiments.

minimize the effect of background noise (M.A. Md Yunus 2010). The total impedance is used as the characterization parameter for the sensor. The electrical equivalent circuit of the series combination sensor is shown in figure 2. The sensor is connected to the output of the signal generator, R_1 denotes the series surface mount resistor connected to the sensor as shown in figure 2(d).

2.2 Experimental setup

The experimental setup is shown in figure 3. The setup has a box containing the waveform generator circuit, signal conditioning circuit, phase and gain measurement circuit, frequency to voltage circuit and temperature circuit. The output signals from the circuits then captured the C8051F020 microcontroller which placed above the box. The microcontroller was interfaced with LabView through RS232 where the impedance calculation (real part, R_{total} and imaginary part, X_{total}), graphical display, and data storage are completed. The temperature sensor, LM335 was placed in the space created between the beaker and polystyrene. The next section will discuss the signal generator circuit, signal condition circuits, phase and gain measurement circuit, impedance (real part, R_{total} and imaginary part, X_{total}) and interface between C8051F020 microcontroller and LabView.

3 THE LOW COST SYSTEM

The low cost system comprises of five main parts: (a) waveform generator circuit, (b) signal conditioning circuit, (c) phase and gain measurement circuit, (d) frequency to voltage circuit, (e) temperature circuit and (f) data acquisition system. The data acquisition system can be broken down into two significant components: (a) C8051F020 microcontroller and (b) LabView program. Next section will discuss about the low cost system in detail.

3.1 Waveform generator circuit

The signal generator described here is based on high XR2206 IC (E. Inc. 1997) and the circuit schematic is illustrated in figure 4. It provides sine around $1Hz$ to $1MHz$, depending on the timing capacitors connected between pin 5 and pin 6. The amplitude can be varied from $1Volt$ peak-to-peak to $12Volt$ peak-to-peak. Potentiometers R1p and R3 can be used to adjust the amplitude and frequency, respectively. The potentiometer, RA, adjusts the sine-shaping resistor, and RB provides the fine adjustment for the waveform symmetry.

3.2 Signal conditioning circuit

The output signal from the waveform generator was connected to a buffer as can be seen in the right circuit of figure 4 and the resulting signal, InputSensor from the buffer is then fed into the sensor (figure 2). LF412 amplifier is chosen as voltage follower, which play a role in the impedance matching, reducing the loading effect as well as protection of the follow-up circuit.

At $10Volt$ peak to peak, $v_1(t)/InputSensor$ gives maximum rating of $23.48dBm$ in a 56Ω system and $v_3(t)/Outputsensor$ amplitude is always less than the maximum amplitude of $v_1(t)/InputSensor$. The signals of $v_1(t)/InputSensor$ and $v_3(t)/Outputsensor$ will be used to calculate the gain and phase which later to be calculate the impedance. However, the phase and gain chip (AD8302) for gain and phase measurement, only cater signals between $-60dBm$ and $0dBm$. Therefore, a pi network attenuator with attenuation level at $50dBm$ was used to attenuate $v_1(t)/InputSensor$ and $v_3(t)/Outputsensor$ as can be seen in figure 5.

3.3 Main description of AD8302

The gain and phase measurements were realized using AD8302 chip (A.D. Inc. 2002). It comprises a closely matched pair of demodulating logarithmic amplifiers where the difference of

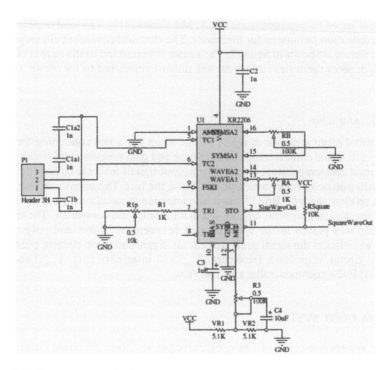

Figure 4. Waveform generator circuit.

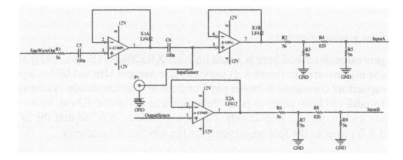

Figure 5. Buffer and attenuator of the input and output signals.

their outputs, a measurement of the magnitude ratio or gain between the two input signals is available and a phase detector of the multiplier type. The indication of working system is provided by a precision $1.8Volt$ reference voltage output. The scanning range of applied input signal is from $-60dBm$ to $0dBm$ in a 50Ω system. The outputs provide an accurate measurement of gain over a $\pm30dB$ range scaled to $30mV/dB$, and of phase over a 0–$180°$ range scaled to $10mV/degree$. Both gain and phase dc outputs should not be over $1.8Volt$. Table 2 summarizes the important information from the datasheet. The circuit of AD8302 can be seen in figure 6 (Semiconductor 1993).

3.4 *Measurement of impedance*

The output signal from the waveform generator was connected to a buffer as can be seen in the circuit of figure 5 and the resulting signal, InputSensor from the buffer is then fed into the sensor (figure 2). The output voltage of the sensor, OutputSensor across the resistor, R_1 is

Table 2. Specification of the AD8302.

Parameters	Condition	Min	Typical	Max	Unit
Input frequency range		≥ 0		≤ 2.7	GHz
Input voltage range	Re = 50Ω	−60		0	dBm
Output voltage minimum	Phase different 180°		30		mV
Output voltage maximum	Phase different 0°		1.8		V
Slew rate			25		V/μs
Output bandwidth			30		MHz
1.8 V reference output	Load = 2kΩ	1.7	1.8	1.9	V
	$V_s = 2.7\text{–}5.5\,V$		0.25		mV/V

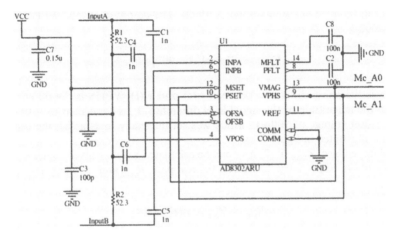

Figure 6.   Connection of AD8302 in gain and phase measurement mode.

connected to a buffer system as can be seen in figure 5. The real part, R_{total} and the imaginary part, X_{total} of the sensor as can be seen in figure 2 can be calculated from:

$$I_1 = \frac{V_3 \angle 0°}{R_1} \tag{1}$$

Where I_1 is the rms value of current through the sensor V_3 is the rms voltage across the surface mount resistor. V_3 is considered as the reference so that its phase angle is 0°. R_1, with nominal value of $120k\Omega$ is the series surface mount resistor used to measure the total current through the sensor. The total impedance Z_{total} is given by,

$$Z_{total} = \frac{V_1 \angle \theta}{I_1 \angle 0} = \frac{V_1}{V_3} \angle \theta_1 \times R_1 \tag{2}$$

Where V_1 and are the rms values of input voltage, $v_1(t)$ and θ_1 is the phase difference between $v_1(t)$ and $v_3(t)$ in degree, taking $v_3(t)$ as reference. The value from the terms V_1/V_3 and θ_1 are given by V_{mag} and V_{phase}, respectively, of AD8302.

At 10Volt peak to peak, $v_1(t)$/InputSensor gives maximum rating of 23.48dBm in a 56Ω system and $v_3(t)$/Outputsensor amplitude is always less than the maximum amplitude of $v_1(t)$/InputSensor. Since the range of input signal power of AD8302 is −60dBm to 0dBm, both $v_1(t)$/InputSensor and $v_3(t)$/Outputsensor signals (after the buffers) were both attenuated by 30dBm using pi attenuators as shown in figure 5. The attenuated signals of $v_1(t)$/InputSensor and $v_3(t)$/Outputsensor are INPA and INPB, respectively. They were then inputted to the AD8302's pins of 2 and 6, respectively as illustrated in figure 6.

For the gain measurement, the AD8302 takes the difference between the output two identical log amplifiers, each taking signals of the same frequency but at different amplitude values. Since the input signals ratio involved frequency domain, the resulting output of MSET pin becomes (A.D. Inc. 2002):

$$V_{MAG} = V_{SLP} \log\left(\frac{V_1}{V_3}\right) + V_{CP} \tag{3}$$

Where V_1 and V_3 are the input signals, V_{MAG} is the dc output corresponding to the magnitude of the signal level difference. VSLP is the slope which is equivalent to $30mV/dB$. VCP is the offset voltage of $900mV$ which was set internally to establish the center point. Equation 3 is illustrated in figure 7(a) (Semiconductor 1993). The dc phase output, VPHS is given by (A.D. Inc. 2002):

$$V_{PHS} = -V_\phi \log\left(|\theta| - 90°\right) + V_{CP} \tag{4}$$

Where θ_1 is the phase difference between $v_1(t)$ and $v_3(t)$, VPHS is the dc output corresponding to the magnitude of the signal level difference. V_θ is equivalent to $-10mV/degree$. VCP is the offset voltage of $900mV$ which was set internally to establish the center point. Equation 4 is illustrated in figure 7(b).

The total impedance, Z_{total} in (2) now can be calculated with the given values of V_1/V_3 and θ_1 from (3) and (4), respectively. Therefore, the real part, R_{total} and the imaginary part, X_{total} are given by:

$$R_{total} = Z_{total} \cos(\theta_1) - R_1 \tag{5}$$

$$X_{total} = Z_{total} \sin(\theta_1) \tag{6}$$

3.5 *Frequency to voltage converter circuit*

In order to display value the frequency of the waveform in section 3.1, VFC110 chip was used. It has high maximum operating frequency of $4MHz$ and a precision $5Volt$ reference. For frequency-to-voltage conversion, SineWaveOut signal from the waveform generator circuit in figure 4 was tapped by frequency to voltage circuit as can be seen in figure 7(a). The signal has to be converted into pulse signal (B.-B. Inc. 1993), before that, SineWaveOut signal was buffered as can be seen in figure 7(a) and LM319N was used to convert the output signal from the buffer as shown in figure 7(b). The resulting pulse signal was applied to the comparator input in figure 7(c).

VFC110 operates by averaging (filtering) the reference current pulses triggered on every falling edge at the frequency input. Voltage ripple with a frequency equal to the input will be present in the output voltage. The magnitude of this ripple voltage is inversely proportional to the integrator capacitor, C10 as shown in figure 7(c). The output signal was limited between 0 and $2.4Volt$ using a differential amplifier circuit as can be seen in figure 7(d). Mc_A6 in figure 7(d) denotes the dc output signal in Volt.

A calibration was performed by adjusting SineWaveOut and observing the frequency using an oscilloscope in the same time Mc_A6 value was jotted down as measured by a digital multimeter. Figure 8 illustrates the frequency value versus Mc_A6 value and the relationship is given by the following equation.

$$\begin{aligned} Frequency\ (Hz) = &-0,071(MC\_A6)^5 + 0,890(MC\_A6)^4 - 2,249(MC\_A6)^3 \\ &+ 2,381(MC\_A6)^2 - 0,922(MC\_A6) + 0,284 \end{aligned} \tag{7}$$

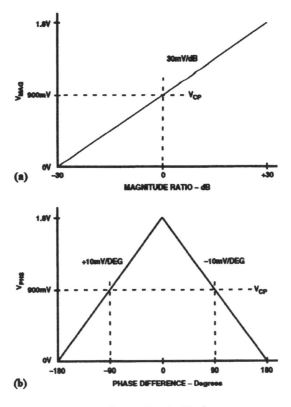

(a)

(b)

Figure 7. Idealized transfer characteristic for the (a) gain (b) phase.

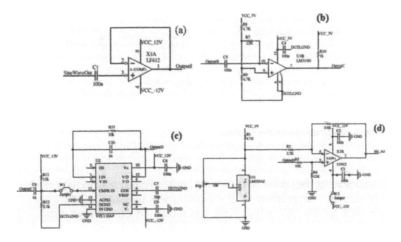

Figure 8. Frequency to voltage converter circuits.

3.6 *Temperature circuit*

Figure 9 depicts the circuit for temperature measurement. It consist of LM335 (Semiconductor 1993) which is used as a sensor to estimate the temperature of the solution. It also consists of a differential amplifier circuit to limit the signal between 0 to 2.4*Volts* as the input of the data acquisition cannot exceed 2.4*Volts*. The calibration to determine the relationship of Mc_A7 output signal with the temperature was established by filling a 250*ml* beaker with

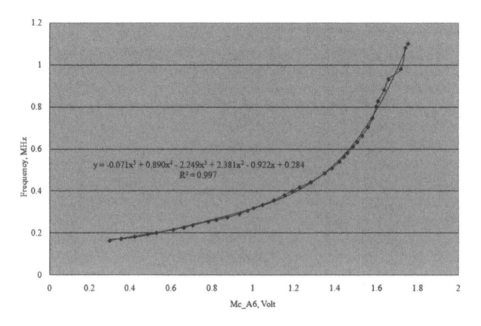

Figure 9. Calibration graph of frequency, MHz versus Mc_A6, *volt*.

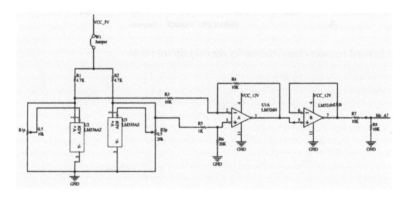

Figure 10. Temperature measurement circuit.

warm water and placing the sensor to the outside of the beaker wall as near as possible. In the calibration process the temperature of the solution was measure using a typical laboratory thermometer while the Mc_A7 output value was measured using a digital multimeter.

$$Temperature \ (celcius) = 218.9(Mc\_A7) - 223.6 \qquad (8)$$

3.7 *Data acquisition system*

C8051F020 microcontroller is used to capture the output signals. The microcontroller has eight channels ADC with 12-bit resolution. Just four channels were used, ADC.0 and ADC.1 were connected to pin 13 (V_{MAG}) and pin 9 (V_{PHS}), respectively of AD8302 chip as shown in figure 6. ADC.6 and ADC.7 were connected to temperature circuit output and frequency to voltage circuit output, respectively. All the channels maximum input level was set to 2.4*Volt*.

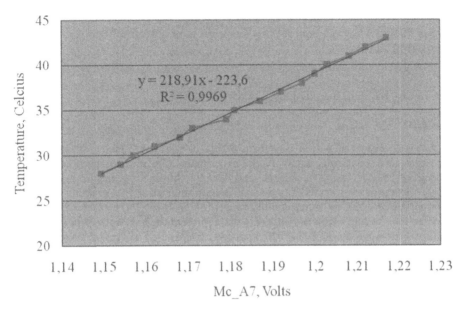

Figure 11.   Calibration graph of solution temperature, Celsius versus Mc_A7, volt.

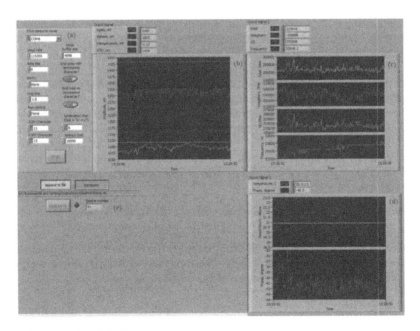

Figure 12.   Frontpanel of LabView program.

The channels are sequentially scanned and the results printed to a PC terminal program via the RS232.

LabView was chosen to calculate (R_{total} and imaginary part, X_{total}), graphical display, and data storage. The front panel of the LabView program can be seen from figure 7. The LabView program comprises of five parts: (a) data transfer configuration (b) analog inputs display (V_{MAG}, V_{PHS}, V_{TEMP}, and V_{F2V}): the results were displayed in an array of string of characters. The output for each channel was obtained and separated from the array and converted from string into equivalent decimal value (c) R_{total}, X_{total}, $|Z|$, and θ_1 display which values have been realized using (1) until (6), (d) temperature and frequency display, and (e) data storage control buttons.

4 METHOD, RESULTS AND DISCUSSIONS

4.1 Method

Each sensor is connected with a waveform generator circuit producing alternating $10Volt$ peak-to-peak sine waveform signals. In the experiment, four measurements were made at frequencies of $164kHz$, $260kHz$, $350kHz$ and $800kHz$. At any frequency setting, the output signals of the sensor are then fed into a signal conditioning circuits containing buffer and pi attenuator circuits. After the signal conditioning, the signals enter a circuit based on AD8302 chip to measure the output signals (in *volts*) that represent the gain and phase values. A data acquisition system based on C8051F020 microcontroller is used to capture the output signals from AD8302 chip, frequency-to-voltage converter, and temperature circuit. The data transfer of the output signals was realized through RS232 into virtual instrument LabView. The impedance calculation (real part (R_{total}) and imaginary part (X_{total})), graphical display, and data storage are completed by LabView where 30 data points were obtained every time the measurement (at any frequency setting) was made.

Every sensor was tested with mili-q water and two lots of aqueous solutions of nitrates forms namely, sodium nitrates ($NaNO_3$) and ammonium nitrates (NH_4NO_3), each of different concentration between $5mg$ and $20mg$ dissolved in $1litre$ of mili-q water. The measurement of the real part, R_{total} and imaginary part, X_{total} values for every water solution type was made at four frequencies, $164kHz$, $260kHz$, $350kHz$, and $800kHz$. The results were then presented in the perspective of impedance spectroscopy analysis approach.

In the next step, for every sensor, a semi-empirical equation based on multilinear regression model was determined from the percentage changes of real part and imaginary part values as compared to the mean real part, R_{total} and imaginary part, X_{total} calculated from mili-q water at the frequencies $260kHz$ and $800kHz$ and estimated concentration (EC) values for every aqueous solution type (sodium nitrates ($NaNO_3$) or ammonium nitrates (NH_4NO_3)). Based on the semi-empirical equations, the performance of the sensors were evaluated and compared to determine the best sensor to detect nitrate contamination in natural water sources.

4.2 Results analysis via electrochemical spectroscopy approach

In this section, electrochemical impedance spectroscopy approach was used to estimate qualitatively the reaction of the sensor with water samples. Furthermore, the response in the frequency domain of an electrochemical reaction can be used to estimate the contamination in the water. Figure 13(a) describes a system for coated electrodes response to aqueous solution as has been described in (Y.J. Tan 1996)/(C.J. McNamara 2004). The circuit composed of the solution resistance (R_s), coating resistance (R_c), coating capacitance (C_c), charge transfer resistance (R_{ct}), and double layer capacitance (C_{dl}). In this research, an additional meander inductance (L) has been added to represent the system as shown in figure 13(b).

A typical electrochemical spectroscopy response with aqueous solution often show both capacitive and inductive semicircle (M. Itagaki 2002). At the system's operating frequency between $164kHz$ and $800kHz$. In general, all the sensors are showing the trait of giving capacitive loop respectively. However, there are significant deviation of the loops as can be seen in figure 15, this is caused by the presence of inductive characteristic (Y.J. Tan 1996) of the meander sensor. As a whole, the impedance curve is shrinking and smaller as the concentration is increased as can be clearly seen from the response of SCS1, SCS3, and SCS4 in figures 15(a), (b), (e), (f), (g), (h). SCS2 was constructed with four number of meander loop seems to have inconsistent response as can be seen in figures (b) and (c).

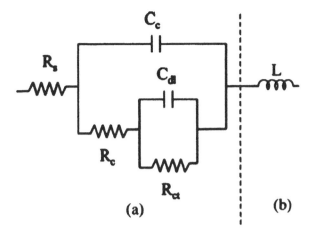

(a)   (b)

Figure 13.   Equivalent circuit to describe the system.

For quantitative results, the real part sensitivities of the sensors are calculated from the mean real part as taking the value from the mean distilled value as reference in the following equation:

$$\sum(\%Real\_part) = \sum_f \left[\frac{(R_{total})_{sample} - (R_{total})_{distilled}}{(R_{total})_{distilled}} \times 100 \right] \qquad (9)$$

The imaginary part sensitivity is calculated from the following equation:

$$\sum(\%Imaginary\_part) = \sum_f \left[\frac{(X_{total})_{sample} - (X_{total})_{distilled}}{(X_{total})_{distilled}} \times 100 \right] \qquad (10)$$

Where (R_{total}) distilled is the mean real part of the impedance value when the sensor is immersed in the distilled water and (R_{total}) sample is the mean real part of the impedance value when the sensor is immersed in the water sample. (X_{total}) distilled is the mean imaginary part of the impedance value when the sensor is immersed in the distilled water and (X_{total}) sample is the mean imaginary part of the impedance value when the sensor is immersed in the water sample. The symbol f represents the frequency ($164kHz$, $260kHz$, $350kHz$ and $800kHz$).

Figures 14 shows the sensitivities values when tested with solution based on $NaNO_3$ and NH_4NO_3. Similar response can be observed for both solution types. In general, for all sensors the real part negatives values progressively decreases with the total concentration of the chemicals with slightly different slope between $NaNO_3$ and NH_4NO_3 plots. This indicates that the electrical conductivity, σ of the water has increased with the increase of concentration of chemical. Good linear correlation of SCS1, SCS3 and SCS3 (as summarized in table 3) between the real part sensitivities with the chemical concentration of $NaNO_3$ and NH_4NO_3, respectively. It is also shown in table 3 that the slope values of the $NaNO_3$ and NH_4NO_3 linear regressions of real part are significantly different for SCS1, SCS3, and SCS3 where the slope of $NaNO_3$ linear regressions is less steep than the slope of $NaNO_3$ linear regressions NH_4NO_3.

The sensitivity of the imaginary part for both type of aqueous solution is increasing when the concentrations were linearly increased in figure 14. The imaginary sensitivities of SCS1, SCS3, and SCS4 give good linear correlation as can be observed in table 3. Double layer capacitance, C_{dl} (figure 15) is formed as ions from the solution accumulate around the electrodes surface (Grahame 1947). The value of C_{dl} is usually higher than the coating, C_c (Y.J. Tan 1996)/(C.J. McNamara 2004), so the small change of C_{dl} will be apparent in the imaginary

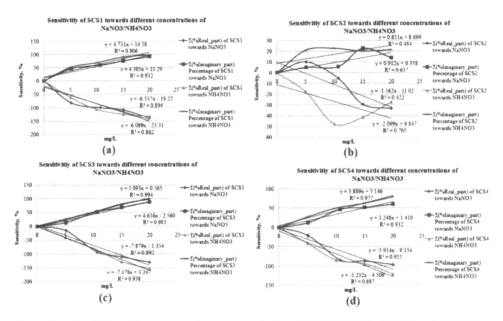

Figure 14. Sensitivity of the sensors to the aqueous solution: (a) SCS, (b) SCS2, (c) SCS3 and (d) SCS4.

Table 3. Summary of linear plots in figure 13.

| Sensor | Linear regression of Σ(%Real_part) | | | | Linear regression of Σ(%Imaginary_part) | | | |
| | $NaNO_3$ | | NH_4NO_3 | | $NaNO_3$ | | NH_4NO_3 | |
	R2	Slope	R2	slope	R2	slope	R2	slope
SCS1	0.862	−6.089	0.894	−6.537	0.932	4.305	0.906	4.731
SCS2	0.765	−2.099	0.422	−1.562	0.637	0.912	0.484	0.851
SCS3	0.938	−7.178	0.992	−7.879	0.983	4.616	0.994	5.093
SCS4	0.887	−5.232	0.955	−5.914	0.932	3.248	0.977	3.889

part of the impedance response. The value of the double layer capacitance depends on many variables such as temperature, ionic concentrations, types of ions, oxide layers, electrode roughness, dielectric properties/relative permittivity (C. Jeyaprabha 2006)/(R. Ravichandran 2005), etc. The change of the imaginary sensitivity is caused by the decrease in C_{dl}, resulting from a decrease in the relative permittivity of the water samples.

It is also shown in table 3 that the slope values of the $NaNO_3$ and NH_4NO_3 linear regressions of imaginary part are significantly different for SCS1, SCS3, and SCS3 where the slope of $NaNO_3$ linear regressions is less steep than the slope of $NaNO_3$ linear regressions NH_4NO_3.

4.3 Multilinear regression model

The semi-empirical equation based on multilinear regression model is determined through the percentage of changes of Real part, R_{total} and Imaginary, X_{total} as compared to the mean R_{total} and Imaginary, X_{total} calculated from mili-q water given as following:

$$\%R_{total\_sample\_f} = \sum \frac{R_{total\_sample\_f} - \overline{R_{total\_miliq\_f}}}{R_{total\_miliq\_f}} \times 100 \tag{11}$$

96

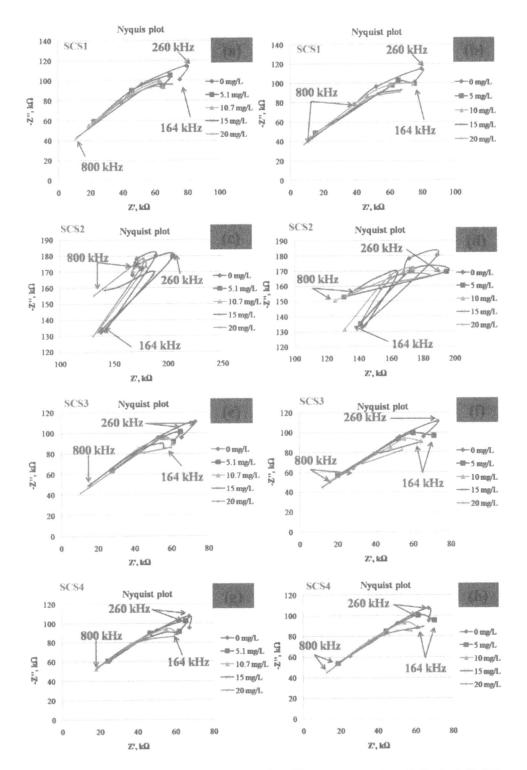

Figure 15. Nyquist plots for experiment involving different concentration of dissolved $NaNO_3$: (a) SCS1 (c) SCS2 (e) SCS3 (g) SCS4. Nyquist plot for experiment involving different concentration of dissolved $NH_4 NO_3$: (b) SCS1 (d) SCS2 (f) SCS3 (h) SCS4.

$$\%X_{total\_sample\_f} = \sum \frac{X_{total\_sample\_f} - \overline{X_{total\_miliq\_f}}}{X_{total\_miliq\_f}} \times 100 \tag{12}$$

Where $\overline{R_{total\_miliq\_f}}$ is the mean value taken from real part of the impedance values when the sensor is immersed in the distilled water at a certain frequency. $R_{total\_sample\_f}$ is the real part of the impedance value when the sensor is immersed in the water sample (mili-q, $NaNO_3$ or NH_4NO_3) at the same frequency as in $R_{total\_miliq\_f}$.

$X_{total\_miliq\_f}$ is the mean value taken form imaginary part of the impedance values when the sensor is immersed in the distilled water at a certain frequency. is the imaginary part of the impedance value when the sensor is immersed in the water sample (mili-q, $NaNO_3$ or NH_4NO_3) at the same frequency as in $X_{total\_miliq\_f}$. The symbol f represents the frequency. From the percentage changes of real part and imaginary part values at $260kHz$ and $800kHz$ and estimated concentration (EC) values in mg/L involving mili-q water and sodium nitrate aqueous solutions. The semi-empirical equations for SCS1, SCS2, SCS3, and SCS4 developed by the multiliner regression (MLR) model using Excel were respectively given as following:

$$SCS1(mg/L_{NaNO_3}) = 0.8508 \times \%R_{total\_sample\_f1} + 0.4501 \times \%R_{total\_sample\_f2}$$
$$- 2.5636 \times \%X_{total\_sample\_f1} - 0.577 \times \%X_{total\_sample\_f2} - 0.1041 \tag{13}$$

$$SCS2(mg/L_{NaNO_3}) = 0.2816 \times \%R_{total\_sample\_f1} - 3.363 \times \%R_{total\_sample\_f2}$$
$$-0.247 \times \%X_{total\_sample\_f1} + 6.8563 \times \%X_{total\_sample\_f2} + 1.3984 \tag{14}$$

$$SCS3(mg/L_{NaNO_3}) = 0.448 \times \%R_{total\_sample\_f1} - 0.024 \times \%R_{total\_sample\_f2}$$
$$- 1.102 \times \%X_{total\_sample\_f1} - 0.1062 \times \%X_{total\_sample\_f2} + 0.091 \tag{15}$$

$$SCS4(mg/L_{NaNO_3}) = 0.6615 \times \%R_{total\_sample\_f1} + 0.5091 \times \%R_{total\_sample\_f2}$$
$$- 1.854 \times \%X_{total\_sample\_f1} - 1.0498 \times \%X_{total\_sample\_f2} + 0.0028 \tag{16}$$

Where and are the percentage changes of the real part and imaginary part, respectively at $260kHz$. $\%R_{total\_sample\_f}$ and $\%X_{total\_sample\_f}$ are the percentage changes of the real part and imaginary part, respectively at $800kHz$. From the percentage changes of real part and imaginary part values at $260kHz$ and $800kHz$ and estimated concentration (EC) values in mg/L involving mili-q water and ammonium nitrate aqueous solutions. The semi-empirical equations for SCS1, SCS2, SCS3, and SCS4 developed by the multiliner regression (MLR) model using Excel were respectively given as following:

$$SCS1(mg/L_{NH_4NO_3}) = 0.0847 \times \%R_{total\_sample\_f1} + 1.263 \times \%R_{total\_sample\_f2}$$
$$- 0.362 \times \%X_{total\_sample\_f1} - 2.14 \times \%X_{total\_sample\_f2} - 0.025 \tag{17}$$

$$SCS2(mg/L_{NH_4NO_3}) = -0.1707 \times \%R_{total\_sample\_f1} - 2.3309 \times \%R_{total\_sample\_f2}$$
$$- 0.9122 \times \%X_{total\_sample\_f1} + 6.7926 \times \%X_{total\_sample\_f2} - 0.1808 \tag{18}$$

$$SCS3(mg/L_{NH_4NO_3}) = 0.3049 \times \%R_{total\_sample\_f1} - 0.068 \times \%R_{total\_sample\_f2}$$
$$- 0.832 \times \%X_{total\_sample\_f1} - 0.023 \times \%X_{total\_sample\_f2} - 0.083 \tag{19}$$

$$SCS4(mg/L_{NH_4NO_3}) = 0.2986 \times \%R_{total\_sample\_f1} + 0.2916 \times \%R_{total\_sample\_f2}$$
$$- 1.026 \times \%X_{total\_sample\_f1} - 0.783 \times \%X_{total\_sample\_f2} - 0.239 \tag{20}$$

For all models in between 13 and 20, all independent values is highly correlated with the dependent value as the R square and Adjusted R Square values are near to 1 as can be seen in table 4 and table 5. The standard error values in table 4 and table 5 represent the estimation of the standard deviation. For all models, the standard deviation value is between 1.8 to 0.40.

By making a new measurement of percentage changes of real part and imaginary part values at $260 kHz$ and $800 kHz$ and involving mili-q water and sodium nitrate aqueous solutions for every sensor, the MLR models in 13, 14, 15, and 16 were tested. Figure 14 shows the plots of estimated concentration (EC) of sodium nitrate against predicted concentration (PC) of sodium nitrate as calculate in 13, 14, 15, and 16. The validation graph involving MLR model of 16 and involving SCS4 had the best R^2 of 0.941, showing that the EC is considerably linear with PC. Table 6 summarizes the EC value and mean PC values (of 30 samples) at each concentration value calculated by MLR models in 13, 14, 15, and 16. In this part, MLR model of

Table 4. Summary of mg/L_{NaNO_3} MLR model.

Model	R^2	Adjusted R^2	Standard error
SCS1(mg/L_{NaNO_3})	0.9747	0.9740	1.1424
SCS2(mg/L_{NaNO_3})	0.9581	0.9570	1.4698
SCS3(mg/L_{NaNO_3})	0.9960	0.9959	0.4518
SCS4(mg/L_{NaNO_3})	0.9918	0.9916	0.6496

Table 5. Summary of $mg/L_{NH_4NO_3}$ MLR model.

Model	R^2	Adjusted R^2	Standard error
SCS1($mg/L_{NH_4NO_3}$)	0.9968	0.9967	0.4098
SCS2($mg/L_{NH_4NO_3}$)	0.9439	0.9423	1.7038
SCS3($mg/L_{NH_4NO_3}$)	0.9900	0.9897	0.7191
SCS4($mg/L_{NH_4NO_3}$)	0.9852	0.9848	0.8746

Table 6. Summary of comparison between EC and PC values involving ammonium nitrate aqueous solutions.

Type of sensor	EC (mg/L)	PC (mg/L)	Standard deviation	Difference (mg/L)
SCS1	0.00	0.8028	0.4542	−0.8028
	5.10	3.9110	0.4228	1.1890
	10.70	6.5016	0.9021	4.1984
	15.00	14.0473	0.5719	0.9527
	20.00	13.9063	0.7923	6.0937
SCS2	0.00	2.4949	2.3158	−2.4949
	5.10	33.0538	2.4794	−27.9538
	10.70	15.9855	1.5185	−5.2855
	15.00	93.2137	2.8495	−78.2137
	20.00	127.8397	3.7672	−107.8397
SCS3	0.00	0.2776	0.2834	−0.2776
	5.10	3.6015	0.7090	1.4985
	10.70	4.4381	0.5396	6.2619
	15.00	9.8086	0.1705	5.1914
	20.00	20.1147	0.4071	−0.1147
SCS4	0.00	0.3232	0.2332	−0.3232
	5.10	3.7324	0.7894	1.3676
	10.70	8.4653	0.7534	2.2347
	15.00	12.5288	0.4235	2.4712
	20.00	22.2579	0.7780	−2.2579

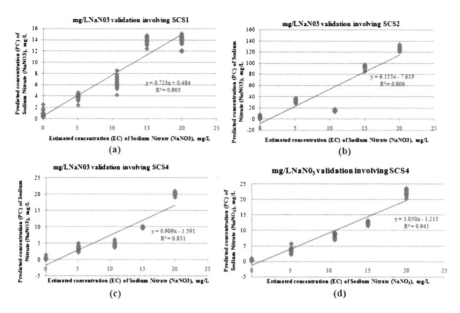

Figure 16. Comparison between EC and PC values involving sodium nitrate aqueous solutions (a) SCS1, (b) SCS2, (c) SCS3, and (d) SCS4.

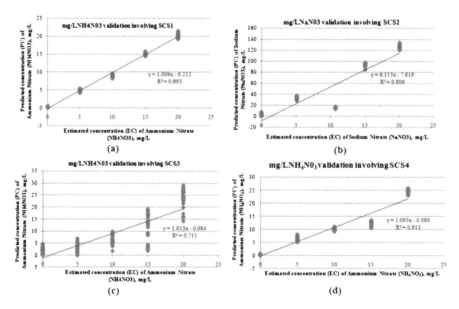

Figure 17. Comparison between EC and PC values involving sodium nitrate aqueous solutions (a) SCS1, (b) SCS2, (c) SCS3, and (d) SCS4.

16 corresponding to SCS1 is the best set can be used in estimating contamination of nitrate in the form of sodium nitrate. The standard deviation, and the average difference between EC and PC, the PC standard deviation is just between 0.25 and 0.79 while the difference is low at low concentration but getting slightly increases as the concentration is increasing between the level of the concentration.

A new measurement of percentage changes of real part and imaginary part values at $260kHz$ and $800kHz$ and involving mili-q water and ammonium nitrate aqueous solutions

Table 7. Summary of comparison between EC and PC values involving ammonium nitrate aqueous.

Type of sensor	EC (*mg/L*)	PC (*mg/L*)	Standard deviation	Difference (*mg/L*)
SCS1	0.00	0.2109	0.1429	−0.2109
	5.00	4.7702	0.3716	0.2298
	10.00	9.0916	0.4396	0.9084
	15.00	15.0102	0.3817	−0.0102
	20.00	20.3282	0.6110	−0.3282
SCS2	0.00	1.8694	1.8459	−1.8694
	5.00	1.9524	1.4417	3.0476
	10.00	2.7130	2.1248	7.2870
	15.00	17.0995	3.7913	−2.0995
	20.00	8.8270	3.4446	11.1730
SCS3	0.00	1.9608	1.3194	−1.9608
	5.00	3.0946	1.9692	1.9054
	10.00	6.5508	1.9412	3.4492
	15.00	10.8344	5.9531	4.1656
	20.00	23.4597	4.0659	−3.4597
SCS4	0.00	0.284328	0.19356	−0.28433
	5.00	6.591721	0.890511	−1.59172
	10.00	10.23552	0.515953	−0.23552
	15.00	12.38866	0.957514	2.611338
	20.00	24.72441	0.701277	−4.72441

was made for every sensor to test the model in 17, 18, 19, and 20. Figure 17 shows the plots of estimated concentration (EC) of ammonium nitrate against predicted concentration (PC) of sodium nitrate as calculate in 17, 18, 19, and 20. The validation graph involving MLR model of 17 and involving SCS4 had the best R^2 of 0.993, showing that the EC is considerably linear with PC as can be seen in figure 17(a). In similar fashion as the previous analysis, table 7 summarizes the EC value andmean PC values (of 30 samples) at each concentration value calculated by MLR models in 17, 18, 19, and 20. As comparison between the results of the sensors in table 7, it can be understood that MLR model of 17 corresponding to SCS1 is the best and can be used in estimating contamination of nitrate in form the form of ammonium nitrate. The standard deviation, and the average difference between EC and PC, the PC standard deviation is just between 0.14 and 0.62 while the difference is low between the level of concentrations.

5 CONCLUSION

This report has successfully demonstrated the capabilities of a low cost system based on planar electromagnetic sensor to be used in estimating the amount of nitrate contamination in water sources. The present study shows that the use of analyzing and modelling the response of planar electromagnetic sensors through electrochemical impedance spectroscopy seem to be very useful in understanding the mechanism behind the interaction between the sensor with the water sample. Furthermore, the validations process of MLR models that has been established for every sensor in estimating nitrates in the form of sodium nitrate and ammonium nitrate have come to a conclusion that the best sensor for estimating sodium nitrate and ammonium nitrate are SCS4 and SCS1, respectively. The use of both sensors have the potential to be used as a useful tool for water sources monitoring in water sources such as river, pond, lake where the nitrate level should not exceed 10 *mg/L*. Moreover, this sensing technique is cost effective, safe to use and non destructive in nature. The proposed sensor system would be able to contribute to the maintaining of the quality of drinking water by continuous monitoring the quality of the drinking water.

REFERENCES

A.D. Inc., A.D. (2002). Available: http://www.analog.com/.

B.-B. Inc., V.D. (1993). Available: http://www.burrbrown.info/.

Bendikov, T.A. and Harmon, T.C. (2005). A sensitive nitrate ion-selective electrode from a pencil lead. An analytical laboratory experiment. *Journal of Chemical Education vol. 82*, pp. 439–441.

Cho, S.-J., e. a. (2002). A simple nitrate sensor system using titanium trichloride and an ammonium electrode. *Sensors and Actuators B: Chemical vol. 85*, pp. 120–125.

Connolly, D. and Paull, B. (2001). Rapid determination of nitrate and nitrite in drinking water samples using ion-interaction liquid chromatography. *Analytica Chimica Acta vol. 441*, pp. 53–62.

D.A.D.B.N. Semiconductor (1993).

Di, H.J., and Cameron, K.C. (2007). Nitrate leaching losses and pasture yields as affected by different rates of animal urine nitrogen returns and application of a nitrification inhibitor—a lysimeter study. *Nutrient Cycling in Agroecosystems vol. 79*, pp. 281–290.

E. Inc., X.-D. (1997). Available: http://http://www.exar.com/.

Ferree, M.A. and Shannon, R.D. (2001). Evaluation of a second derivative uv/visible spectroscopy technique for nitrate and total nitrogen analysis of waste water samples. *Water Research vol. 35*, pp. 327–332.

Goldfine, N.J. (1993). Magnetometers for improved material characterization in aerospace application. *Material Evaluation*, pp. 396–405.

Goldfine, N.J., Clark, D. and Lvett, T. (1995). Material characterization using model based meandering winding eddy current testing (mw-et). *EPRI Topical Workshop: Electromagnetic NDE Applications in the Electric Power Industry*.

Grahame, D.C. (1947). The electrical double layer and the theory of electrocapillarity. *Chemical Reviews vol. 41*, pp. 441–501.

Gumede, N.J. (2008). Harmonization of internal quality tasks in analytical laboratories case studies: water analysis methods using polarographic and voltammetric techniques. Master's thesis, Faculty of Applied Sciences, Durban University of Technology, Durban.

Itagaki, M., e. a. (2002). Deviations of capacitive and inductive loops in the electrochemical impedance of a dissolving iron electrode. *Analytical Sciences vol. 18*, pp. 641–644.

Jeyaprabha, C., e. a. (2006). Corrosion inhibition of iron in 0.5 mol · l^{-1} H_2SO_4 by halide ions. *Journal of the Brazilian Chemical Society vol. 17*, pp. 61–67.

Karunanayaka, D., Gooneratne, C.P.M.S.C. and S.G.G. (2006). A planar electromagnetic sensors aided non-destructive testing of currency coins. *NDT.net vol. 11*, pp. 1–12.

Kjr, T., e. a. (1999). Sensitivity control of ion-selective biosensors by electrophoretically mediated analyte transport. *Analytica Chimica Acta vol. 391*, pp. 57–63.

McNamara, C.J., e. a. (2004). Biodeterioration of incralac used for the protection of bronze monuments. *Journal of Cultural Heritage vol. 5*, pp. 361–364.

Md Yunus, M.A., e. a. (2010). A new planar electromagnetic sensor for quality monitoring of water from natural sources. *ICST 2010. 4th International Conference on Sensing Technology*, pp. 554–559.

Metcalf, N.F., e. a. (1987). The differing sensitivities of the hemoglobin of fetal and adult red-cells to oxidation by nitrites in man—the role of plasma. *Proceedings of the Physiological Society*, pp. 44. Cambridge Meeting: Poster Communications.

Moorcroft, M.J., e. a. (2001). Detection and determination of nitrate and nitrite: a review. *Talanta vol. 54*, pp. 785–803.

Mukhopadhyay, S.C. (2002). Quality inspection of electroplated materials using planar type micro-magnetic sensors with post-processing from neural network model. *IEEE Proceedings-Science Measurement and Technology vol. 149*, pp. 165–171.

Ravichandran, R., e. a. (2005). Corrosion inhibition of brass by benzotriazole derivatives in nacl solution. *ROYAUME-UNI: Emerald vol. 52. Bradford*.

Stanley, E.M. (2009). *Fundamental of Environmental Chemistry* (3rd ed.). Taylor and Francis Group.

Tan, Y.J., e. a. (1996). An investigation of the formation and destruction of corrosion inhibitor films using electrochemical impedance spectroscopy (eis). *Corrosion Science vol. 38*, pp. 1545–1561.

Terblanche, A.P.S. (1991). Health hazards of nitrate in drinking water. *Water Sa vol. 17*, pp. 77–82.

Lecture Notes on Impedance Spectroscopy – Kanoun (ed)
© *2012 Taylor & Francis Group, London, ISBN 978-0-415-69838-2*

Impedance spectroscopy of novel interdigital sensors for endotoxin detection

A.R. Mohd Syaifudin, S.C. Mukhopadhyay & P.L. Yu
School of Engineering and Advanced Technology, Massey University, Palmerston North, New Zealand

Michael J. Haji-Sheikh
College of Engineering and Engineering Technology, Northern Illinois University, DeKalb, IL, USA

Cheng-Hsin Chuang
Department of Mechanical Engineering, Southern Taiwan University, Tainan, Taiwan

ABSTRACT: Novel interdigitals sensors have been fabricated to detect the presence of Lipopolysaccrides (LPS) which is related to endotoxin contamination in food. Three different configurations of electrode structure with different substrates and fabrication methods have been developed and analysed. The impedance spectroscopy (IS) has been used to analyse the sensors' characteristic as well as to analyse the interaction between sensors with different LPS solutions. Pure LPS products of O111:B4 and LPS rough strain have been used in the experiments and it was found that the impedance spectroscopy of the developed sensors able to show good experimental results. This paper reports the successful development of novel planar interdigital sensors and the impedance spectroscopy measurement for different LPS.

Keywords: LPS, interdigital, FR4, alumina, glass, impedance, capacitance

1 INTRODUCTION

Endotoxins (Lipopolysaccharide, LPS) are dangerous toxins which are associated with Gram-negative bacteria (S.A. Seydel U 2000). Endotoxins or LPS are part of the outer membrane of the Gram-negative bacteria cell wall which comprises of O antigen, Core and Lipid A (S. Hauschildt and Muller-Loennies 2000). The detection of endotoxins which are caused by the pathogens such as Escherichia coli and Salmonella is the main concerns the food industries (O. Lazcka and Munoz 2007). The conventional method of detection is using culturing techniques. It is conducted by enriching the food sample and performing various media-based metabolic testing (agar plates or slants) (Leoni and Legnani 2001). Conventional techniques are capable of giving accurate result because of high selectivity and sensitivity. The conventional method of detection are tedious, time-consuming and expensive (Alocilja and Radke 2003). Polymerase chain reaction (PCR) is one of the most popular methods of pathogens detection (Fratamico 2003). PCR is based on the isolation, amplification and quantification of a short DNA sequence. Recent advancement in PCR technology has developed a real-time PCR (D. Rodriguez-Lazaro and Ikonomopoulos 2005) and a multiplex PCR (A. Jofre and Aymerich 2005) which can obtain results in a few hours. Enzyme linked immunosorbent assays (ELISA) and culture techniques for determining and quantifying pathogens in food have been established (McMeekin 2003). ELISA is combined with a specific antibodies and the sensitivity of simple enzyme assays by using antibodies or antigens coupled to an easily assayed enzyme (O. Lazcka and Munoz 2007). The technique is developed to replace the detection using culture and colony counting. However, the ELISA also needs to be confirmed using conventional test. Biosensor has been widely used to detect pathogens in food (O. Lazcka and Munoz 2007), (Alocilja and Radke 2003). Biosensor is one of the rapid methods, based on immunochemical or nuclei acid technologies (O. Lazcka and Munoz 2007). The transduction methods used in biosensor are optical fibre (Cambridge 2003), (Cooper 2003), electrochemical (Mubammad-Tahir and Alocilja 2003), interdigitated array microelectrodes (Varshney and Li 2009), surface plasmon resonance (SPR) (Morris and Sadana 2005) and others.

Planar interdigital sensors or also known as interdigitated electrode structures have been used in biosensor applications (Varshney and Li 2009), (Radke and Alocilja 2005). Planar interdigital sensors are used for characterizing of near-surface properties such as conductivity, permeability and dielectric properties. We have been using these planar sensors in estimation of properties of dielectric material for milk, and saxophone reeds (S. C. Mukhopadhyay and Demidenko 2006), (S.C. Mukhopadhyay and Deidenko 2007), characterization of different materials of complex permittivity (E. Fratticcioli and Sorrentino 2004), estimation of ship skin (S.C. Mukhopadhyay and Norris 2008) and detection of contaminated seafood with dangerous toxin (A.R.M. Syaifudin and Mukhopadhyay 2009). The combination of planar interdigital sensors or interdigitated array electrode with impedance spectroscopy measurements have been reported in (J. Ramon-Azcon and Marco 2008), (E. Koep and Liu 2006). The impedance spectroscopy (IS) is a powerful technique used to evaluate electrical properties of materials and their interfaces with surface-modified electrodes (Barsoukov and Macdonald 2005). Recent method of detection in pathogen sensing has been discussed by Heo et al, (Heo and Hua 2009) which utilized the applications of interdigital sensors with impedance spectroscopy measurement. The electrical properties for detection of pathogenic bacteria also has been reported in (M.S. Mannoor and McAlpine 2010). This paper reports the development of novel planar interdigital sensors and their impedance spectroscopy measurements for endotoxins detection.

2 NOVEL PLANAR INTERDIGITAL SENSOR

The interdigital sensor operates as a parallel plate capacitor. The electric field lines generated by the sensor will penetrate through the material under test (MUT) and interact with it (A.V. Mamishev and Zahn 2004). The sensor behaves as a capacitor in which the capacitance becomes a function of system properties. By measuring this capacitance the system properties can be evaluated (Mukhopadhyay and Gooneratne 2007). The capacitance between a positive and negative electrode can be measured by;

$$C = \frac{\varepsilon_0 \varepsilon_r A}{d} \tag{1}$$

where C is the capacitance, ε_0 is the permittivity of free space $(8.854 \times 10 - 12)$, ε_r is the relative permittivity, A is the effective area and d, is effective spacing between positive and negative electrodes. In order to get a strong signal, the electrode pattern are repeated many times (K. Sundara-Rajan and Mamishev 2004). Electric field distribution between positive and negative electrodes can have multiple excitation patterns at different level of proximity for different electrode arrangements with suitable pitch length (distance between two adjacent electrodes). The penetration depth of the designed sensor can be calculated from two adjacent electrodes of similar polarity (A.V. Mamishev and Zahn 2004). Based on these information novel interdigital sensors have been developed to have optimum numbers of negative electrodes, higher penetration depth and uniform electric field distribution throughout the sensor geometry. The optimum number of negative electrodes between two positive electrodes of interdigitated configuration contributes to highest sensitivity measurement.

The sensor impedance can be calculated by equation 2 where, V_e is the excitation voltage with small AC amplitudes, the sensing voltage, V_s across the series resistance (R_s) is observed, to measure the current (I) flowing to the sensor.

$$Z = \frac{V_e}{V_s} R_s \qquad (2)$$

The selection of the R_s should be such that it does not influence the impedance of the sensor and at the same time sufficient signal is available across it for the purpose of measurement. The chosen R_s ($120k\Omega$) produced a good sensing voltageas well as given better phase angle which is close to $90°$. Both the magnitude and the phase of the sensor impedance need to be measured. The absolute value of the real part of the sensor (R) and the imaginary part (capacitive reactance, X_c) is given by;

$$R = |Z|\cos\theta - R_s \qquad (3)$$

$$X_c = |Z|\sin\theta \qquad (4)$$

The phase angle, θ measured was closed to $90°$, so the real part of the sensor impedance, R is very small compared to the imaginary part. Also the change of resistance with the dielectric material is negligible. The real part has not been considered for estimation of the system properties. Therefore, at low operating frequency, the capacitive reactance becomes the only parameter being measured in the system. The effective capacitance can be calculated by;

$$C = \frac{1}{2\pi f X_c} \qquad (5)$$

3 SENSOR DESIGN AND CONFIGURATION

Three sensors with different electrode configurations have been designed. The configurations were based on different number of negative electrodes introduced in the sensors design. All sensors have same effective area of $4750\mu m$ by $5000\mu m$ and having pitches of $250\mu m$. The positive and negative electrodes have the same length and width of $4750\mu m$ and $125\mu m$ respectively. All sensors were designed using Altium Designer 6 software. The representations of novel interdigital sensors are shown in Figure 2. Table 1 shows the parameters used to design the sensors with different configurations. A collaborative work has been established with Northern Illinois University, United States and Southern Taiwan University, Taiwan for sensors fabrication. The sensors were designed first at Massey University and then were sent to those universities for different method of fabrication.

4 SENSOR SUBSTRATE AND FABRICATION

The sensors were fabricated on three different materials with different fabrication process. Different substrates were used for different fabrication process will have different influence on

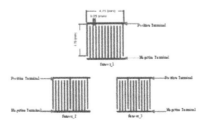

Figure 1. Representation of novel interdigital sensors with configuration #1 (Sensor_1), configuration #2 (Sensor_2) and configuration #3 (Sensor_3).

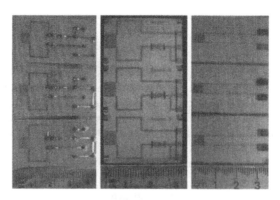

Figure 2. The fabricated novel interdigital sensors on different substrates (a) Fiberglass—FR4 (b) Alumina (c) Glass.

Table 1. Novel interdigital sensor parameters for different configurations.

Sensor configuration	Sensing area (mm^2)	Pitch length (mm)	Number of electrodes	
			Positive	Negative
Sensor_1–11	23.75	0.250	2	11
Sensor_1–5	23.75	0.250	3	10
Sensor_1–3	23.75	0.250	4	9

the sensitivity of the fabricated sensors (J. Hong and No 2005). Figure 2 shows the fabricated interdigital sensors on different substrates. The first design of sensors was fabricated using photographic printed circuit board technology and was fabricated on the fiberglass, FR4. The sensor design was printed on Overhead Projector (OHP) transparency/film and placed onto the pre-coated photoresist FR4 board. The board was then exposed to the Ultraviolet (UV). The UV light transmits through the clear portion of the film and cures the photoresist. After that, the board was submerged into a developer bath which will develop and removes the sensitized photoresist. The neat traces of the interdigital electrode structures are being constructed on the FR4 board. The second group of sensors was fabricated using alumina as a substrate. The sensors were constructed using standard thick film printing methodologies. The pattern was drawn in AutoCad and then printed on a Fire 9500 photoplotter. The patterns were then transferred to three 325 mesh screens. The printer used in the fabrication was a Presco 435, which is capable of printing up to a $100mm \times 100mm$ substrate. The top trace layer was printed using a PdAg $850°C$ firing alloy while the ground plane was made using a $850°C$ silver material. The solder dam was made using a $600°C$ low thermal conductivity dielectric. The wet paste was dried and fired after each individual layer was printed. The paste was dried at $150°C$ then inspected. The dried substrate was then placed on the belt of a BTU firing furnace profiled to deliver $850°C \pm 5°C$ for 10 minutes. The entire firing cycle is approximately 30 minutes. The substrates were then patterned for the next layer and the cycle was repeated. The final layer is a low thermal conductivity dielectric to limit the solder flow during the mounting of a chip resistor. The third design of sensors was fabricated on a $25 \times 75mm^2$ microscope slide by conventional micromachining techniques. The glass slide was cleaned in the acetone solution for five minutes and followed by a five minutes ultrasonic bath in a methanol solution. The cleaned glass then was dried using with an N2 gun and dehydrated by a hotplate for an additional 30 minutes at $225°C$. After the cleaning, $300Å$ of chromium (Cr) and $700Å$ of gold (Au) were deposited in sequence on the surface of glass slide by E-beam evaporator. The interdigital electrodes (IDT) were patterned by photolithography process using a positive photoresist (Shipley 1813). The developed photoresist was utilized to be an etching mask for the wet etching of metal layer. Finally, the sensor was obtained after removal of the photoresist.

5 MEASUREMENT AND APPARATUS

The impedance spectroscopy was utilized to characterize the sensors characteristic without material under test and also to evaluate the sensors for different LPS structure. The measurements were performed using MT 4090 LCR Meter with basic accuracy of 0.1%. The LCR measurement mode was selected for impedance measurement, phase angle, serial and parallel capacitance. Each measurement was recorded for different test frequencies of $100Hz$, $120Hz$, $1kHz$, $10kHz$, $100kHz$ and $200kHz$ with the test voltage was set to $1Vrms$. The measurements were carried out as shown in Figure 3. The pH readings of each solution were measured using pH meter model 420A from Orion Research Inc.

6 EXPERIMENTAL RESULTS AND DISCUSSION

6.1 *Sensor characteristic analysis*

Each fabricated sensor has different characteristic which shows different sensing performance. The impedance spectroscopy measurement was used to analyse each sensor for their characteristic. The Nyquist plot of each sensor is shown in Figure 4 for different sensors with different configurations. Figure 5 shows the impedance characteristic of all sensors for different configurations. Both figures have shown that sensor fabricated with FR4 has very high impedance as compared to sensor fabricated on alumina and glass. The FR4 sensors are expected to have low sensing performance as compared to others because of these characteristic. The capacitance characteristic is shown in Figure 6. The result clearly shown that sensor fabricated on Alumina and Glass has higher capacitance value which indicates that they will have better sensing performance. The phase angle measurement of each sensor is shown in Figure 7. It is observed that at low frequency, the phase angle of each sensor is closed to 90° where they behave as pure capacitive sensors but as the frequency increases the phase angle

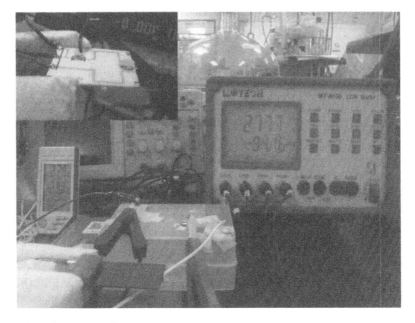

Figure 3. Experiment setup for sensor measurement.

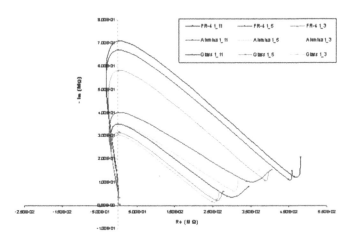

Figure 4.   Nyquist plot showing sensors' characteristics of different configurations.

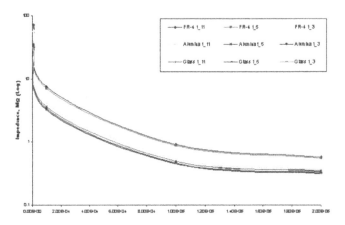

Figure 5.   Impedance characteristic of different configurations ($1kHz$–$200kHz$).

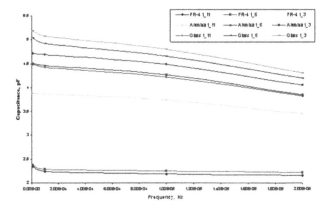

Figure 6.   Capacitance for different configurations ($1kHz$–$200kHz$).

also change. At high frequency the sensors lead to a parasitic resistance or ESR (Equivalent Series Resistance) and at higher frequency theybecome inductive. Therefore, at low frequency ($\leq 200kHz$) the planar interdigital sensor has the ability to evaluate the dielectric measurement of material under test.

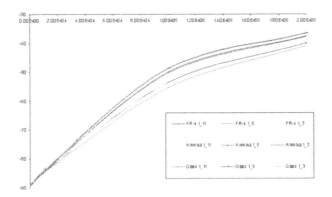

Figure 7. Phase measurements of different configuration ($1kHz$–$200kHz$).

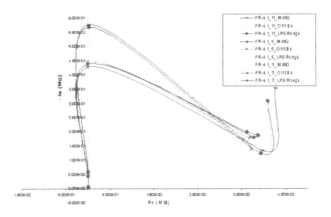

Figure 8. Nyquist plot of FR4 sensors for different LPS structures.

6.2 *Sensors test with LPS solution*

Experiments were conducted to evaluate the Lipopolysaccharides (LPS) structure from LPS rough and LPS O111:B4. The LPS were purchased from Sigma-Aldrich, United States. Sigma L2630 (O111:B4) is the wild type LPS and Sigma L9641 is from the rough strain (mutant without the O-specific chain, also made of polysaccharide and a target for specific antibody). These LPS were extracted using phenol extraction which contains less than 3% of protein (Catalogue 2010). $1mg$ of each sample of O111:B4 and LPS rough were diluted into $1ml$ of Milli-Q water. The pH reading of each solution was measured. It was observer that the pH reading of Milli-Q water was 6.50, pH of O111:B4 solution is 6.59 and pH of LPS rough solution is 6.33. A thin glass slide of $150\mu m$ was placed on the sensors to avoid direct contact between the electrodes and the sample. A small amount ($40\mu l$) of each solution was pipetted using $10\mu l$ pipettor on the glass slide. The impedance measurements and phase angle were recorded before and after the solution were pipetted on the glass slide. The results of different sensors were presented using Nyquist plots showing the impedance spectra of Milli-Q water (control), LPS rough and O111:B4. Figures 8, 9 and 10 show the impedance spectra of each solution using FR4 sensors, Alumina sensors and Glass sensors respectively. Results show that sensors with eleven numbers of negative electrodes (1–11) have better performance and can clearly discriminate between each solution. Results in Figure 11 show that Alumina sensor (1–11) and Glass sensor (1–11) appear to have higher sensitivity measurement compared to FR4 sensor. It is noticed that the change in absolute value of the real part of the signal is very small (negligible). So the effect of conductivity does not have any influence on the signal.

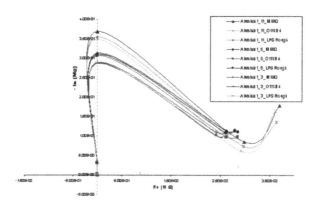

Figure 9.　Nyquist plot of Alumina sensors for different LPS structures.

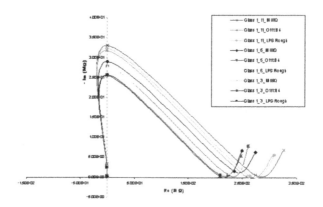

Figure 10.　Nyquist plot of Glass sensors for different LPS structures.

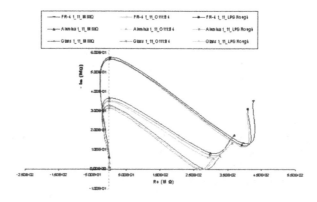

Figure 11.　Nyquist plot of Milli-Q water (control), O111:B4 and LPS rough.

Therefore, the change of measured signal is due to the change of the effective permittivity of the solutions. The fabricated novel planar interdigital sensors are able to characterize the dielectric spectroscopy of different LPS structure. It can be said that these sensors can be used to detect the presence of endotoxins. Further experiments are currently conducted to investigate the significance of any molecular selectivity or resolution as well as the effect of residual adsorption.

7 CONCLUSION

New types of interdigital sensors have been fabricated to investigate their sensing performance. New sensors have been fabricated using different method of fabrication and were fabricated on different substrates. Due to different values of permittivity of different substrate, the capacitance values and corresponding impedance values are influenced by these substrates. Impedance spectroscopy has been utilized to characterize the fabricated sensors. Experiments were conducted to investigate how sensors behave with pure LPS which are related to endotoxins. Results have shown that sensor with eleven number of negative electrodes (1–11) has better sensitivity measurement. More experiments are currently conducted to immobilize specific binding property of LPS (Polymyxin B) for specific molecular selectivity and to enhance the sensor performance.

ACKNOWLEDGEMENTS

The authors would like to acknowledge Massey University, New Zealand, Northern Illinois University, United States, Southern Taiwan University, Taiwan, and Malaysian Agricultural Research and Development Institute (MARDI), researchers referenced throughout the paper and also to whom that had fruitful discussions and collaboration with the authors.

REFERENCES

Alocilja, E. and Radke, S. (2003). Market analysis of biosensor for food safety. *Biosensors & Bioelectronics vol. 18*, pp. 841–846.

Barsoukov, E. and Macdonald, J.R. (2005). Impedance spectroscopy: Theory, experiment, and applications. *Second (ed) New Jersey: Wiley Interscience.*

Cambridge, E.W.P.L. (2003). Biosensors for environmental pollutants and food contaminants. *Analytical and Bioanalytical Chemistry vol. 377*, pp. 434–445.

Catalogue, S.-A. (2009–2010). Biochemicals, reagents and kits for life science research. pp. 1437–1440.

Cooper, M.A. (2003). Label-free screening of bio-molecular interactions. *Analytical and Bioanalytical Chemistry vol. 377*, pp. 834–842.

Fratamico, P.M. (2003). Comparison of culture, polymerase chain reaction (pcr), taqman salmonella, and transia card salmonella assays for detection of salmonella spp. in naturally-contaminated ground chicken, ground turkey, and ground beef. *Molecular and Cellular Probes vol. 17*, pp. 215–221.

Fratticcioli, E., M.D. and Sorrentino, R. (2004). A simple and low-cost measurement system for the complex permittivity characterization of materials. *IEEE Transactions on Instrumentation and Measurement vol. 53*, pp. 1071–1077.

Hauschildt, S., Brabetz, W., A.B.S.L.H.P.Z.E.T.R. and Muller-Loennies, S. (2000). Structure and activity of endotoxins. *Handbook of experimental pharmacology vol. 145*, pp. 619–667.

Heo, J. and Hua, S.Z. (2009). An overview of recent strategies in pathogen sensing. *Sensors vol. 9*, pp. 4483–4502.

Hong, J., Yoon, D.S., S.K.K.T.S.K.S.K.E.Y.P. and No, K. (2005). Ac frequency characteristics of coplanar impedance sensors as design parameters. *Lab on a Chip vol. 5*, pp. 270–279.

Jofre, A., Martin, B., M.G.M.H.M.P.D.R.-L. and Aymerich, T. (2005). Simultaneous detection of listeria monocytogenes and salmonella by multiplex pcr in cooked ham. *Food Microbiology vol. 22*, pp. 109–115.

Koep, E., Jin, C.M., M.H.R.D.R.N.K.S.R.S. and Liu, M.L. (2006). Microstructure and electrochemical properties of cathode materials for sofcs prepared via pulsed laser deposition. *Journal of Power Sources vol. 161*, pp. 250–255.

Lazcka, O., F.J.D.C. and Munoz, F.X. (2007). Pathogen detection: A perspective of traditional methods and biosensors. *Biosensors & Bioelectronics vol. 22*, pp. 1205–1217.

Leoni, E. and Legnani, P.P. (2001). Comparison of selective procedures for isolation and enumeration of legionella species from hot water systems. *Journal of Applied Microbiology vol. 90*, pp. 27–33.

Mamishev, A.V., Sundara-Rajan, K., F.Y.Y.Q.D. and Zahn, M. (2004). Interdigital sensors and transducers. *Proceedings of the IEEE vol. 92*, pp. 808–845.

Mannoor, M.S., Zhang, S.Y., A.J.L. and McAlpine, M.C. (2010). Electrical detection of pathogenic bacteria via immobilized antimicrobial peptides. *Proceedings of the National Academy of Sciences of the United States of America vol. 107*, pp. 19207–19212.

McMeekin, T.A. (2003). Detecting pathogens in food. *Cambridge, England: Woodhead Publishing Limited vol. 1.*

Morris, B.A. and Sadana, A. (2005). A fractal analysis for the binding of riboflavin binding protein to riboflavin immobilized on a spr biosensor. *Sensors and Actuators B-Chemical vol. 106*, pp. 498–505.

Mubammad-Tahir, Z. and Alocilja, E.C. (2003). A conductometric biosensor for biosecurity. *Biosensors & Bioelectronics vol. 18*, pp. 813–819.

Mukhopadhyay, S.C. and Gooneratne, C.P. (2007). A novel planar-type biosensor for noninvasive meat inspection. *IEEE Sensors Journal vol. 7*, pp. 1340–1346.

Mukhopadhyay, S.C., Choudhury, S.D., T.A.V.K. and Norris, G.E. (2008). Assessment of pelt quality in leather making using a novel non-invasive sensing approach. *Journal of Biochemical and Biophysical Methods vol. 70*, pp. 809–815.

Mukhopadhyay, S.C., Goonerate, C., G.S.G. and Demidenko, S. (2006). A low cost sensing system for quality of dairy products. *IEEE Transactions on Instrumentation and Measurements vol. 55*, pp. 1331–1338.

Mukhopadhyay, S.C., Woolley, J.D.M., G.S.G. and Deidenko, S. (2007). Saxophone reed inspection employing planar electromagnetic sensors. *IEEE Transactions on Instrumentation and Measurements vol. 56*, pp. 2492–2503.

Radke, S. and Alocilja, E. (2005). A microfabricated biosensor for detecting foodborne bioterrorism agents. *IEEE Sensors Journal vol. 5*, pp. 744–750.

Ramon-Azcon, J., Valera, E., A.R.A.B.B.A.F.S.-B. and Marco, M.P. (2008). An impedimetric immunosensor based on interdigitated microelectrodes (id mu e) for the determination of atrazine residues in food samples. *Biosensors & Bioelectronics vol. 23*, pp. 1367–1373.

Rodriguez-Lazaro, D., D'Agostino, M., A.H.M.P.N.C. and Ikonomopoulos, J. (2005). Real-time pcr-based methods for detection of mycobacterium avium subsp paratuberculosis in water and milk. *International Journal of Food Microbiology vol. 101*, pp. 93–104.

Seydel U, S.A., Blunck, R, B.K. (2000). Chemical structure, molecular conformation, and bioactivity of endotoxins. *Chem Immunol vol. 74*, pp. 5–24.

Sundara-Rajan, K., L.B. and Mamishev, A.V. (2004). Moisture content estimation in paper pulp using fringing field impedance spectroscopy. *IEEE Sensors Journal vol. 4*, pp. 378–383.

Syaifudin, A.R.M., K.P.J. and Mukhopadhyay, S.C. (2009). A low cost novel sensing system for detection of dangerous marine biotoxins in seafood. *Sensors and Actuators B-Chemical vol. 137*, pp. 67–75.

Varshney, M. and Li, Y.B. (2009). Interdigitated array microelectrodes based impedance biosensors for detection of bacterial cells. *Biosensors & Bioelectronics vol. 24*, pp. 2951–2960.

Lecture Notes on Impedance Spectroscopy – Kanoun (ed)
© 2012 Taylor & Francis Group, London, ISBN 978-0-415-69838-2

Characterisation of electromechanically active polymers using electrochemical impedance spectroscopy

Indrek Must, Karl Kruusame, Urmas Johanson, Tarmo Tamm,
Andres Punning & Alvo Aabloo
Intelligent Materials and Systems Lab, Institute of Technology, University of Tartu, Tartu, Estonia

ABSTRACT: Characterisation and application of transducers composed of electromechanically active polymer composites can be essentially improved by the electrical modelling of these materials using the electrochemical impedance spectroscopy method. In this article, the concept of using ionic polymer composites as position sensors is characterised using impedance spectra. Two different types of ionic polymers—carbon-polymer composite and ionic polymer-metal composite—are considered. Equivalent circuits of these two materials are specified. Additionally, a method for describing degradation effects of the materials is discussed.

Keywords: CPC, IPMC, position sensor, transducer, electroactive polymer composite

1 INTRODUCTION

Electromechanically Active Polymer (EAP) composites are materials that exhibit changes in their size or shape when stimulated by an electric field. The most common applications of EAPs are actuators and sensors (in which case mechanical manipulation generates electrical response). Possible application fields for EAP transducers include biomedical applications, biomimetic robotic devices, noise damping systems, space and military appliances etc (M. Shahinpoor 2005).

In the current paper, we consider two types of ionic EAPs, which are rather similar in construction and operation, but are composed of different materials and have substantial differences in operating mechanism. The materials considered are Ionic Polymer-Metal Composite (IPMC) and Carbon Polymer Composite (CPC, also PCC). The common quality of the two materials is that they consist of a polymer membrane, the opposite planar surfaces of which are covered with layers of electrically conductive materials. The electric potential between the electrodes causes transverse movement of mobile ions inside the composite, forcing the actuator to bend. These actuators exhibit large actuation at low applied voltage (usually less than $3V$).

These EAPs have also sensorial properties. Mechanical deformation induced by external forces causes transverse movement of ions inside the sheet; that in turn induce changes of electrical properties of different parts of the composite. These changes could be registered as short-circuit current between opposing electrodes or changes in impedance of different parts of the composite.

IPMC is probably the best known type of ionic EAPs. The first publications about this material appeared in early 1990s. IPMC consists of a thin ionic polymer membrane, covered by some inert metal electrodes at the opposite planar sides. Typical ionic polymer (e.g. Nafion) contains suitable mobile counter-ions to balance the charge of the anions fixed to the polymer backbone, and solvent—typically water or ionic liquid. Applied voltage causes migration of cations inside polymer matrix towards cathode, which in turn results in a non-uniform distribution of the ions inside polymer. As a result, the polymer sheet bends towards the electrode with more negative potential. To avoid drying, the water-based IPMC actuators are used in aqueous or humid environment. A comprehensive review of the operating principles of IPMC can be found in (M. Shahinpoor 2001), sensorial properties are thoroughly reviewed in (M. Shainpoor 2001), (D. Pugal 2010). The experiments described in the current paper were conducted with an IPMC consisting of 0.25mm thick Nafion membrane plated with platinum electrodes. The introduced cations were Li+. CPC material has gained a lot of interest and the research in a direction of finding the most perspective materials for this type of composite is of great significance. A CPC actuator is made of two layers of electrodes comprising carbon materials of high specific surface area. The electrodes are separated by a thin ion-permeable polymer film. The whole sandwich contains ionic liquid as electrolyte. The voltage applied between the carbon electrodes causes electro-osmotic effect in the nanoporous material, which results in the migration of the cations to one side of the material, and the anions to the opposite side. This results with unbalanced distribution of ions between the electrodes and—as usually cations are larger than anions—the whole material bends towards the cathode. CPCs are operated in ambient conditions (J. Torop 2010a), (J. Torop 2010b).

In this paper, experiments using CPC materials composed of carbide-derived carbon (CDC), polyvinylidene fluoride (PVDF) polymer and 1-ethyl-3-methlyimidazolium tetrafluroborate (EMIBF4) as the ionic liquid are described. CDC is a nanoporous material with controlled pore size distribution and very high electric double-layer capacitance. The pores in the CPC material can be described as a transmission line with distributed ionic conductance and capacitance. Transmission line behaviour can be observed both between opposing electrodes as well as between opposing ends of each electrode layer.

All EAP materials used were fabricated on-site.

To date, the EIS method has been applied in several papers for the characterisation of the composition and behaviour of IPMCs (K. Kruusame 2009), (S. Leary 1999), (Z. Chen 2008), (K. Takagi 2007), (T. Nakamura 2009), and CPCs (I. Takeuchi 2010), (H. Randriamahazaka 2010), however, the amount of papers regarding this subject remains relatively low.

This paper presents impedance spectra of CPC and IPMC materials and their electrodes. The concept of using these two types of materials as position sensors is verified experimentally using the impedance spectra.

Material degradation effects are important to know for practical applications and are so far not paid much attention to in the literature. We propose a methodology of describing the degradation effects of ionic EAPs with the EIS methods.

2 EXPERIMENTAL

In this work, three different configurations were used for connecting the impedance measurement terminals to the EAP material. Figure 1 represents these configurations employed for measurement of different properties of the devices. An impedance spectrometer Parstat 2273, manufactured by Princeton Applied Research was used in the experiments at frequencies $5mHz$–$10kHz$, and the Precision Impedance Analyser 6500B manufactured by Wayne Kerr Electronics at frequencies $20Hz$–$100kHz$. All measurements were performed with sinusoidal excitation at $10mV$ RMS.

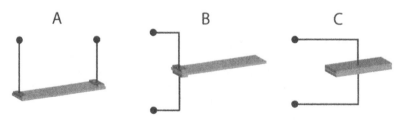

Figure 1. Different setups for impedance measurements. A-Electrode impedance; B-Cross electrode impedance; C-Sheet impedance.

Electrode impedance was measured by placing voltage gold terminals to opposite edges of one electrode layer of an assembled composite actuator (Figure 1: A). The effect of a separator and the opposing electrode layer was considered to be small, as the resistance between opposite edges of an electrode is much smaller than between opposing electrodes. IPMC material was measured in the hydrated state.

Impedance between the electrodes (hereinafter referred as cross-electrode impedance) was measured by placing clamp-shaped voltage terminals on opposing sides of one edge of an actuator (Figure 1: B). It is apparent that the effect of surface impedance also contributes to this value.

Both, electrode and cross-electrode impedance were registered first in lank (free deflection) mode. After that, the composite sheet was bent to opposing sides while trying to retain equal bending curvatures.

In degradation experiments of IPMC material, impedance was registered by placing a polymer sheet completely between voltage terminals of larger surface area than the sheet, as depicted in Figure 1: C.

3 RESULTS

3.1 *EIS measurements*

An impedance plot of a CPC electrode depicted in Figure 2 shows mainly resistive behaviour at high frequencies ($\geq 10kHz$) and more capacitive behaviour at lower frequencies. As expected, the impedance magnitude of an electrode in expanded state i.e. along the convex side of a bent actuator, is more than twice of the magnitude compared to an electrode in lank state. This can be interpreted with the increase of the distance between carbon particles next to each other. However, one can see that the electrode impedance magnitude also rises when the electrode is compressed i.e. along the concave side of an actuator. One would normally expect an increase in conductivity in a compressed electrode as it has been reported in the case of IPMC material (D. Pugal 2010), (A. Punning 2007). Further on, the impedance phase graph shows clearly that the phase on the expanded electrode increases, and on the compressed electrode decreases in almost equal extent. From this data, we can conclude that it is

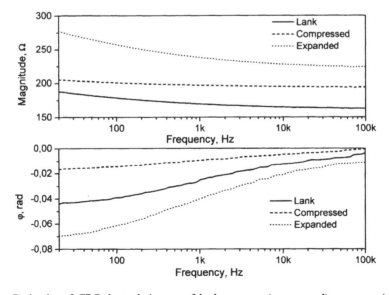

Figure 2. Bode plot of CPC electrode in case of lank, concave (compressed) or convex (expanded) shape.

possible to estimate the deflection of CPC actuator by registering the phase behaviour. The impedance plot of lank IPMC material is presented in Figure 3. Cross-electrode impedance spectrum of CPC depicted in Figure 4 shows a strong increase in magnitude when the sheet is bent. The effect is not dependent on the bending direction. Although the sheet was bent to opposite sides to an equal extent, the magnitude of changes are different. This behaviour is in accordance with the behaviour of surface resistance, and gives an indication of different electrical properties of each electrode. From the phase plot, we can see that the phase declines in equal extent when sheet is bent to opposite sides. The behaviour of the cross-electrode impedance of the IPMC material depicted in Figure 5 shows a strong but almost equal change in impedance magnitude and phase when bent to opposing sides.

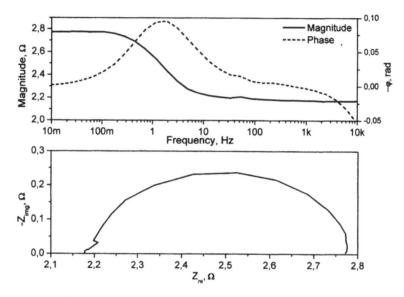

Figure 3.   Bode and Nyquist plots of the lank IPMC electrode.

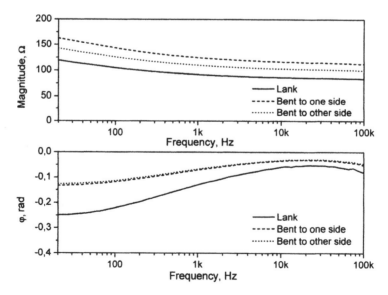

Figure 4.   Bode plot of CPC cross-electrode impedance.

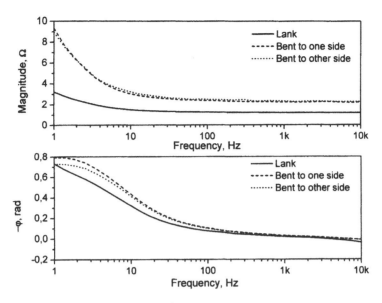

Figure 5. Plot of IPMC cross-electrode impedance.

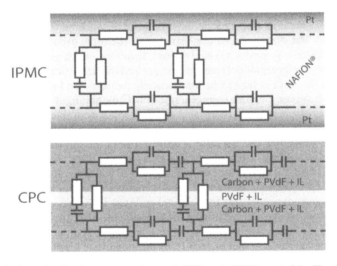

Figure 6. Equivalent circuits for cross-sections of CPC and IPMC materials. IL stands for ionic liquid.

3.2 *Equivalent circuits*

The introductory equivalent circuit of IPMC is comprehensively described in (A. Punning 2009). It consists of infinite series of infinitesimally short single units. Each unit comprises of discrete elements representing the impedance of electrodes and the impedance across the separating polymer layer. This equivalent circuit is eventually comparable to a sophisticated RC transmission line.

According to the measurements described hereinabove, the approach described in (A. Punning 2009) was extended for both materials considered. The circuits depicted in Figure 6 are also composed of distributed elements representing the impedances of the electrodes and the separating polymer layer. The equivalent circuits between surface layers (electrode effects suppressed) of both materials is proved being as described in (A. Punning 2009),

117

i.e. as the capacitance of the double layer coupled with the ionic resistance and ohmic resistance of the electrolyte and solvent.

Table 1 gives rough physical parameter values of the materials, obtained as the result of the experiments.

Additionally, several capacitive components in the electrode layers of both materials were observed. The capacitive component in the electrode layer of IPMC is reported in (K. Kruusame 2009) and in case of CPC in (H. Randriamahazaka 2010). This component can be explained only by knowing the physical structure of the electrodes. An IPMC electrode consists of a layer of discrete metal clusters tens of micrometers in size formed during the manufacturing process. A CPC electrode layer contains rigid carbon particles with diameter around $1 \mu m$. The bending or stretching behaviour of such electrodes differs from that of a homogenous film. The electronic conductivity is constituted by direct contact of the conductive particles touching each other. The capacitive component is presumably originating from the electric double-layer formed between the conductive particles.

In the electrode of CPC material, an extra capacitance not connected in parallel with resistances was observed. This indicates that some particles in the electrode have no electronic contact and they are charged only through the electric field. Therefore, the complete charging process of this type of system takes infinite amount of time and the capacitance can only be given as a function of frequency.

3.3 Degradation effects

Degradation effects were performed on IPMC material only.

When an IPMC material undergoes a large number of working (bending) cycles, various effects can occur. The metal cations associated with the ionic groups in the polymer side-chains can be replaced by protons (i.e. H+ ions), therefore increasing the conductivity, but protons are not able to generate volumetric changes and the deflection extent decreases. Also, the polymer and electrode layer can be mechanically damaged and the material can be contaminated. It is, however, possible to regenerate IPMC by boiling it in HCl solution for the removal of the contamination and after that to reintroduce desirable ions. In order to investigate the degradation effects, sinusoidal excitation signals for a high number of cycles were applied upon an unused IPMC actuator. Figure 7 shows the impedance spectra registered consecutively before cycling, after two cycling intervals and after changing the counter-ions to H+ by boiling in HCl. The actuator was first cycled 20 000 times at $1.8V$ and consecutively 3 600 cycles at $2.8V$. Figure 7 shows that smaller number of cycles at higher voltage causes more changes in impedance spectra than 20 000 cycles at $1.8V$. No changes in the extent of deflection amplitude were observed during the experiment. After the replacement of all cations contained in the composite to protons during boiling in HCl, conductance increases by the order of one magnitude, as can be seen from the magnitude graph in Figure 7 at frequencies above $100 Hz$. One can also observe an increase in conductance after a large number of working cycles.

This experiment shows clearly that higher working voltages lead to faster degradation of this type of material as the electrolysis of water contributes more at higher potentials. Figure 8 presents the voltage-current relation of IPMC material in semilogarithmic scale. At the potential of about $2.3V$, the slope of the graph increases rapidly, giving an evidence

Table 1. Typical values of some physical parameters of IPMC and CPC materials.

	IPMC	CPC
Double layer capacitance	$10mcF/cm^2$	$20mF/cm^2$
Ionic resistance	$1\Omega/cm^2$	$10\Omega/cm^2$
Resistance of the electrolyte or solvent	$5 \cdot 102\Omega/cm^2$	$102..104\Omega/cm^2$
Electrode resistance	$1\Omega/square$	$10..103\Omega/square$

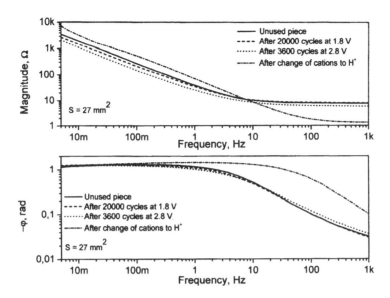

Figure 7.   Degradation of IPMC. Consecutive experiments were performed with a single IPMC actuator.

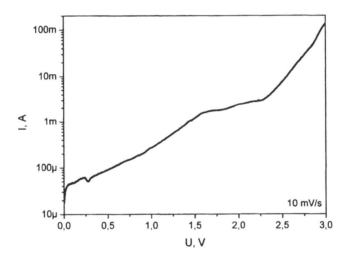

Figure 8.   Voltage-current relationship of an IPMC actuator with water used as solvent.

of more dominant redox processes on electrodes. The electrochemical window of water is known to be $1.23V$. The cause of increase of suitable working potential in case of IPMC material by 0.97 volts is caused by the ionic membrane resistance and the low hydrogen ion concentration inside the ionic polymer membrane.

4   DISCUSSION

The EAPs considered in the current paper can have huge capacitance (up to several dozen mF/cm^2) and moderately high resistance of electrodes. It is obvious that the characteristic area of impedance spectra is located at the very low frequency range, often falling behind millihertz range. The measurements at microhertz range are extremely time consuming and in some cases cannot be carried out, as it is probable that the properties of the system under test will change during the measurement.

To overcome this problem, a straightforward approach is proposed. A (preferably non-inductive) resistor of appropriate resistance is to be added in parallel with the system under study. This will shift the characteristic frequency range of the whole system towards the higher frequencies. Later, the effect of the extra resistor is subtracted from the resulting data. A diversity of equivalent circuits can be parameterized by this method using equivalent impedance conversion formulas found in the literature (Zobel 1923).

5 CONCLUSION

Based on the EIS measurements, we have demonstrated the possibility of using CPC and IPMC materials as position sensors. The external bending of the EAP materials considered causes changes in both electrode and cross-electrode impedances of these materials. The electrode impedance of CPC and IPMC changes differently when the sheets are bent. In the case of CPC, the impedance magnitude increases when the electrode is expanded as well as compressed; while that of IPMC depends on the direction of bending. The degradation experiments with the IPMC material indicate good durability only when excited with voltages not exceeding the electrochemical window of the solvent.

ACKNOWLEDGMENTS

The financial support from Estonian Science Foundation grant no 7811, Targeted financing SF0180008s08 from Estonian Ministry of Education and Estonian Information Technology Foundation is acknowledged.

Help from Prof. Mart Min from Tallinn University of Technology in the field of impedance measurements is kindly appreciated.

REFERENCES

Chen, Z., X.T. (2008). A control-oriented and physics-based model for ionic polymer-metal composite actuators. *IEEE/ASME Transactions on Mechatronics vol. 13, no. 5*, pp. 519–529.

Kruusame, K., Punning, A., M.K.A.A. (2009). Dynamical variation of the impedances of IPMC. *Proc. SPIE vol. 7287*, 72870V.

Leary, S., Y.B.-C. (1999). Electrical impedance of ionic polymeric metal composites. *Proc. SPIE vol. 3669*, pp. 81–86.

Nakamura, T., Ihara, T., T.H.T.M.K.A. (2009). Measurement and modelling of electro-chemical properties of ion polymer metal composite by complex impedance analysis. *SICE Journal of Control, Measurement, and System Integration vol. 2, no. 6*, pp. 373–378.

Pugal, D., Jung, K., A.A.K.K. (2010). Ionic polymer-metal composite mechanoelectrical transduction: review and perspectives. *Polym. Int. vol. 59*, pp. 279–289.

Punning, A., Johanson, U., M.A.A.A.M.K. (2009). A distributed model of ionomeric polymer metal composite. *Journal of Intelligent Material Systems and Structures vol. 20 no. 14*, pp. 1711–1724.

Punning, A., Kruusmaa, M., A.A. (2007). Surface resistance experiments with IPMC sensors and actuators. *Sensor Actuat. A-Phys. vol. 133 no. 1*, pp. 200–209.

Randriamahazaka, H., K.A. (2010). Electromechanical analysis by means of complex capacitance of bucky-gel actuators based on single-walled carbon nanotubes and an ionic liquid. *J. Phys. Chem. C vol. 114*, pp. 17982–17988.

Shahinpoor, M., K.K. (2001). Ionic polymermetal composites: I. fundamentals. *Smart Mater. Struct. vol. 10*, pp. 819–833.

Shahinpoor, M., K.K. (2005). Polymer metal composites: IV. industrial and medical applications. *Smart Mater. Struct. vol. 14*, pp. 197–214.

Takagi, K., Nakabo, Y., Z.L.K.A. (2007). On a distributed parameter model for electrical impedance of ionic polymer. *Proc. SPIE vol. 6524*, 652416.

Takeuchi, I., Asaka, K., K.K.T.S.K.M.H.R. (2010). Electrochemical impedance spectroscopy and electromechanical behavior of bucky-gel actuators containing ionic liquids. *J. Phys. Chem. C vol. 114*, pp. 14627–14634.

Torop, J., Arulepp, M., J.L.A.P.U.J.V.P.A.A. (2010a). Nanoporous carbide-derived carbon material-based linear actuators. *Materials vol. 3*, pp. 9–25.

Torop, J., Kaasik, F., T.S.A.A.K.A. (2010b). Electromechanical characteristics of actuators based on carbide-derived carbon. *Proc. SPIE vol. 7642*, 76422A.

Zobel, O.J. (1923). Theory and design of uniform and composite electric wave filters. *Bell Systems Technical Journal vol. 2*, pp. 1–46.

Lecture Notes on Impedance Spectroscopy – Kanoun (ed)
© 2012 Taylor & Francis Group, London, ISBN 978-0-415-69838-2

Electrochemical impedance spectroscopy of systems with low electrolytic conductivity

Frank Berthold & Winfried Vonau
Kurt-Schwabe-Institut für Mess- und Sensortechnik e.V., Meinsberg, Ziegra-Knobelsdorf, Germany

ABSTRACT: The knowledge of the mechanism of action of influences on the electrochemical system to be investigated is precondition for an optimal interpretation of measuring results in case of practical application of the Electrochemical impedance spectroscopy (EIS). In the limit range of that application, e.g. in case of measurements at systems with low electrolytic conductivity, incorrect results can occur. Taking not process-relevant time constants into account, the analysis of complex models shows the action of impedances influencing the measuring results. It can be resumed that in dynamic investigations of systems with low electrolytic conductivity the design of the measuring station can be optimised knowing and including parasitic components in a model (equivalent circuit).

Keywords: Electrochemical impedance spectroscopy, equivalent circuit, low electrolytic conductivity, measuring cell

1 INTRODUCTION

Electrochemical impedance spectroscopy (EIS) is well established as a method for the characterisation of the condition of phase boundaries since about 30 years. An advantage is that it does not come to irreversible system changes because the stationary equilibrium condition is only marginal disturbed. The knowledge of the mechanism of action of influences on the system to be investigated is precondition for an optimal interpretation of measuring results in case of practical application of the method (Cahan B D 1972), (Ghr H 1984), (Fiaud C 1987). In the limit range of that application, e.g. in case of measurements at systems with low electrolytic conductivity (aprotic solvents), incorrect results can occur. In this contribution the behaviour of 1-Butanol/$0.15 mol \cdot L^{-1}$ $NaClO_4$ is illustrated exemplarily. Taking not process-relevant time constants into account, the analysis of complex models shows the action of impedances influencing the measuring result. Accordingly, the construction of the measuring station can be optimised.

2 EFFECTS OF THE ELEMENTS OF THE EQUIVALENT CIRCUIT

At a specific conductivity of the electrolytic solution of $\leq 1 mS \cdot cm^{-1}$ in a measuring cell according to Fig. 1 capacitances and resistances (that are not a component of the electrode impedance) can disturb the result of the impedance determination in a frequency range between $10 Hz$ and $100 kHz$. Thereby time constants partly cancel out each other. The interpretation of the impedance curve becomes difficult. It appears appropriate to consider different parameters separately in their mode of action. The electrode impedance in Fig. 2 is given by equation 1:

$$Z = \frac{u}{i} = R_O + \frac{R_P}{1 + j\omega R_P C_D} \qquad (1)$$

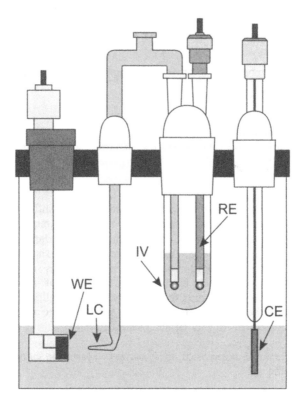

Figure 1. Cell arrangement for the EIS-measurements.
WE: working electrode; RE: reference electrode; CE: counter electrode; LC: luggin capillary; IV: intermediate vessel.

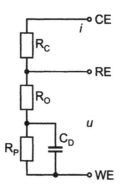

Figure 2. Equivalent circuit.
R_P = polarisation resistance; C_D = double layer capacitance; R_O = resistance of the electrolytic solution between WE and LC; R_C = resistance of the electrolytic solution between LC and CE.

Capacitances between the thin walled Luggin capillary (LC) and the working and counter electrode as well as input and cable capacitances of the potentiostat have certain effects. Starting from equivalent circuit according to Fig. 3 capacitive operating time constants (being caused by capacitors C_L and C_E) in combination with the resistances R_L, R_S and R_E as well as inductive operating time constants (being caused by capacitor C_G) also in combination with the resistances R_L, R_S and R_E can be found. For the calculation of the transfer function

124

$Z = u/i$ with suitable math software programms Fig. 3 is converted, e.g. into an equivalent circuit given in Fig. 4. This leads to equation 2.

$$Z = \frac{(S_2 + Z_A) \cdot Z_E \cdot (R_E + Q_3)}{(S_2 + Z_A + S_3) \cdot (Q_3 + R_3 + Z_E) + Q_3 \cdot (R_3 + Z_E)} \tag{2}$$

The used parameters are defined as follows:

$$Z_D = \frac{1}{j\omega C_D} \quad Z_E = \frac{1}{j\omega C_E} \quad Z_G = \frac{1}{j\omega C_G} \quad Z_L = \frac{1}{j\omega C_L}$$

$$Z_W = R_O + \frac{R_P Z_D}{R_P + Z_D} \quad Z_A = \frac{Z_W Q_1}{Z_W + Q_1} \tag{3}$$

$$Q_1 = \frac{R_L Z_L}{R_S} + R_L + Z_L \quad Q_2 = \frac{R_L R_S}{Z_S} + R_L + R_S \quad Q_3 = \frac{R_S Z_L}{R_L} + R_S + Z_L$$

$$S_1 = R_C + Z_G + Q_2 \quad S_2 = \frac{R_C Q_2}{S_1} \quad S_3 = \frac{Z_G Q_2}{S_1}$$

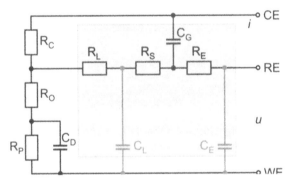

Figure 3. Equivalent circuit with parasitic resistances and capacitances.
R_p = polarisation resistance; C_p = double layer capacitance; R_o = resistance of the electrolytic solution between WE and LC; R_C = resistance of the electrolytic solution between LC and CE; R_L, R_S = resistances of the electrolytic solution in the LC; C_L = stray capacitance between WE and LC; R_E = resistance of the reference electrode; C_E = cable and input capacitance of the potentiostat; C_G = stray capacitance between CE and LC.

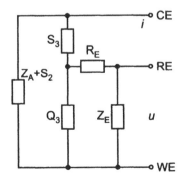

Figure 4. Equivalent circuit for computer calculations generated from Fig. 3. Explanation of symbols see text.

Measured impedance spectra (IM6e—Zahner-Elektrik GmbH) of Mg/1-Butanol/ $0.15 mol \cdot L^{-1}$ $NaClO_4$ (Fig. 5, curve 1) are compared with calculated spectra of the transfer function (2) depending on circuit element values R_L, R_S, R_E, C_L, C_E, C_G (Fig. 5, curves 2–4).

3 INFLUENCE OF THE CONSTRUCTION OF THE MEASURING DEVICE

3.1 *Distance of the LC from the working electrode*

The distance of the LC from the working electrode determines the electrode resistance R_O and the capacitance C_L between LC and working electrode (Fig. 1 and Fig. 3). A variation of this distance leads to a change of the time constant that is formed from R_L, R_S and C_L.

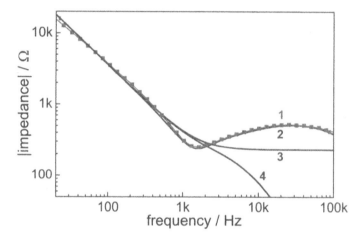

Figure 5.   Impedance spectra Mg/1-Butanol/$0.15 mol \cdot L^{-1}$ $NaClO_4$ and influence of the equivalent circuit parameter on the impedance spectra according to Fig. 3.
Curve 1: measured impedance spectra; Curves 2 to 4: calculated spectra according to Fig. 3; Curves 2 to 4: $R_P = 1.5 M\Omega$, $C_D = 440 nF$, $R_O = 230\Omega$, $R_C = 500\Omega$; Curves 2, 4: $R_L = 150 k\Omega$, $R_S = 300 k\Omega$, $R_E = 50 k\Omega$, $C_L = 160 pF$, $C_E = 50 pF$; Curve 2: $C_G = 100 pF$; Curve 3: $R_L = R_S = R_E = 0$, $C_L = C_E = C_G = 0$; Curve 4: $C_G = 0$.

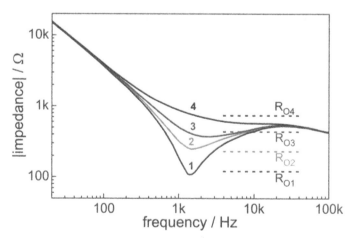

Figure 6.   Influence of the distance a of the Luggin capillary to the working electrode on impedance spectra of the system Mg/1-butanol/$0.15 mol \cdot L^{-1}$ $NaClO_4$.
Curve 1: $a = 0.5 mm$; Curve 2: $a = 1 mm$; Curve 3: $a = 2 mm$; Curve 4: $a = 4 mm$; R_{O1} ... R_{O4}: estimated values of the electrolytic resistance R_O between working electrode and Luggin capillary according to equation (3).

In Fig. 6 the influence of the distance of the LC on the modulus of the impedance is shown. For comparison the estimated values for R_o from equation 3 are plotted with r = radius of the circular working electrode, a = distance between the working electrode and the LC and κ = specific conductivity of the electrolyte solution. For impedance measurements in media with $\leq 1mS \cdot cm^{-1}$ the distance a should be $\geq 2mm$. In polarisation investigations the error which is caused by an $i \cdot R_o$ drop is to be considered.

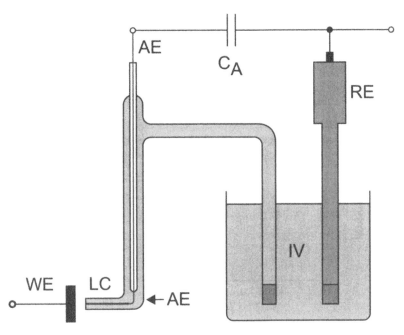

Figure 7.   Capacitive bypass of the Luggin capillary.
WE: working electrode; LC: luggin capillary; AE: platinum auxiliary electrode; CA: auxiliary capacitor; RE: reference electrode; IV: intermediate vessel.

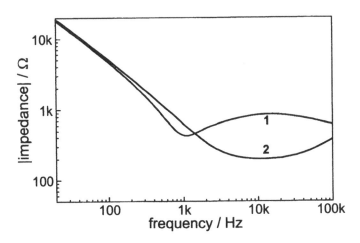

Figure 8.   Effect of the influence of a capacitive bypass of the Luggin capillary on impedance spectra of the system Mg/1-butanol/$0.15 mol \cdot L^{-1}$ $NaClO_4$.
Curve 1: starting condition according to Fig. 1; Curve 2: capacitive bypass according to Fig. 7.

3.2 *Capacitive bypass of the LC*

The arrangement of a platinum auxiliary electrode AE within the LC and the capacitive bypass to the input of the potentiostat (Fig. 7) cause a decrease of the disturbing effect by the electrolytic resistance $(R_L + R_S)$ in the LC and the resistance R_E of the reference electrode (Oelner W 2006). The auxiliary capacitor C_A shortens the entire reference electrode setup at higher frequencies. This perfection can hardly be reached in practice. The disturbed frequency range always is shifted to higher frequencies (Fig. 8).

3.3 *Arrangment of the intermediate vessel*

By an inappropriate arrangement of the intermediate vessel it is possible that it dips into the electrolytic solution. Thereby an additional effective stray capacitance to the counter electrode occurs. Fig. 9 demonstrates the impact on the impedance spectra.

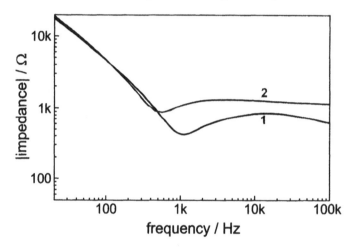

Figure 9. Effect of the arrangement of the intermediate vessel on impedance spectra of the system Mg/1-butanol/0.15$mol \cdot L^{-1}$ $NaClO_4$.
Curve 1: starting condition according to Fig. 1; Curve 2: intermediate vessel dips into the electrolytic solution.

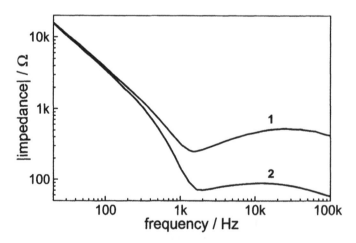

Figure 10. Effect of the influence of the input capacitance of the potentiostat on impedance spectra of the system Mg/1-Butanol/0.15$mol \cdot L^{-1}$ $NaClO_4$.
Curve 1: input amplifier connected directly with the reference electrode; Curve 2: without external input amplifier.

3.4 *Potentiostat input*

The input capacitance C_E of the potentiostat consists of the cable capacitance for the connection of the reference electrode and of the capacitance of the input circuit.

The effective capacitance can be decreased by an input amplifier which optional is placed at the measuring cell. This is possible with the system IM 6e used for the measurements. Fig. 10 demonstrates the influence of input and cable capacitances on the impedance curve.

4 CONCLUSIONS

Dynamic measurements at low electrolytic conductivity should be carried out with specially adapted measuring cells and potentiostates. The measuring cell can be optimised by appropriate construction measures. By means of variation of all model parameters actions of parasitic effects caused by the stray capacitances and the influence of the electrolyte conductivity become clear.

For special applications it is to be checked whether a separation of the reference electrode from the measuring medium is necessary using an intermediate vessel. Stray capacitances can be minimised additionally by appropriate guarding circuit techniques.

REFERENCES

Cahan B.D, Nagy Z, G.M.A. (1972). *J. Electrochem. Soc. 119*, 64.
Fiaud C, Keddam M, K.A.T.H. (1987). *Electrochim. Acta 32*, 445.
Ghr H, Mirnik M, S.C.A. (1984). *J. Electroanal. Chem. 180*, 273.
Oelner W, Berthold F, G.U. (2006). *Mater. Corros. 57*, 455.

Lecture Notes on Impedance Spectroscopy – Kanoun (ed)
© *2012 Taylor & Francis Group, London, ISBN 978-0-415-69838-2*

Recent experiments at CiS (Center of intelligent Sensor technology) concerning CVD/ALD layer monitoring

Ingo Tobehn, Konrad Hasche, Michael Hintz, Stefan Völlmeke & Arndt Steinke
CiS Forschungsinstitut für Mikrosensorik und Photovoltaik, Erfurt, Germany

ABSTRACT: We will establish a new possibility to monitor thin layers to determine their thickness or the relative permittivity as well. For this aim the well established interdigital structures from CiS are modified and enhanced. The focus lies on materials with a very high permittivity (for example a layer with $\varepsilon_r = 80$ (TiO_2 for example)) based on a SiO_2 substrate with $\varepsilon_r = 3.9$, used in semiconductor manufacturing e.g. Knowing ε_r, the thickness of the layer can be determined, or, the other way round, knowing the thickness of the layer, ε_r can be determined.

Keywords: high relative permittivity, IDS, ellipsometry, CVD, ALD, microelectronics

1 INTRODUCTION

After computational simulations and calculations, describing the environment, we will perform first measurement in the next weeks. For this intention, several wafers were prepared by the Fraunhofer IKTS in Dresden with thin layers and relative permittivities from $\varepsilon_r \approx 10$ up to 150. There is a wide range of possible applications; for example in process controlling in semiconductor manufacturing.

2 DESCRIPTION OF THE SETUP

In figure 1, the main part of the equipment is shown. It is the back side of the sensor, consisting of a LTCC substrate with a thickness of about $100\mu m$. The interdigital structure with various layouts is shown in the right picture (S. Voellmeke 2010). It will be placed headfirst on the LTCC substrate in the recessed square. The surface of the sensor must be parallel to the substrate with high accuracy, else it isn't possible to place the whole surface on the sample. The sensor device is mounted on a CNC machine to perform the measurements automatically over the whole surface of a 4" wafer. The configuration of the interdigital structure varies from $2.5\mu m/1.5\mu m$ electrode with/gap up to $200\mu m/20\mu m$.

3 PRINCIPLE OF THE MEASUREMENT

To explain the principle of the method, figure 2 shows an example with a layer, having a relative permittivity ε_r of 80 on a normal silicium wafer ($\varepsilon_r = 11.8$). Depending of the thickness, the sensor detects a relative permittivity in the range between 11.8 and 80. With this measurement, knowing the functional relationship between ε_r and the layer thickness it is easy to determine this value. In the case of a layer thickness of $0\mu m$, the corresponding values are well known; for very high values of the layer thickness, the results corresponds to a sample, consisting only of this material and the silicium wafer delivers no fraction to the result. The penetration depth of the electric field can be varied by using different interdigital structures.

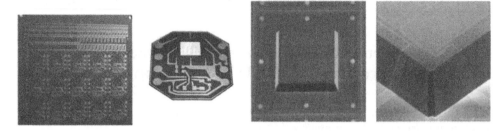

Figure 1. LTCC substrate and the interdigital mesa-structure. The whole equipment will be installed on a CNC machine in combination with a CiS made pressure sensor to detect the resting of the inter-digital structure on the surface.

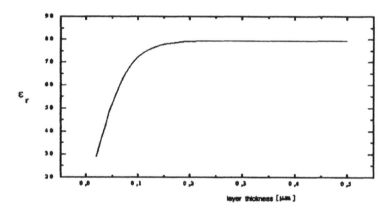

Figure 2. Relative permittivity ε_r as a function of the layer thickness.

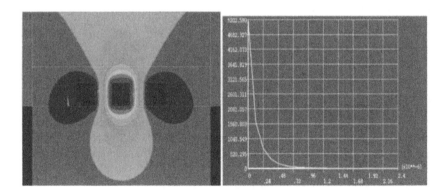

Figure 3. Left: Potential of a structure with 180nm electrode with and 180nm Gap; Right: Norm of the electric field intensity as a function of the depth of penetration.

In figure 3, a simulation with Ansys 11.0 is shown using a structure with 180μm electrode with and 180μm gap.

REFERENCES

Voellmeke, S., K.P.A.S. (2010). Technologiekompatible 3D- Strukturierung zu Herstellung integrierter Mikrosysteme. *Technologien und Werkstoffe der Mikrosystem- und Nanotechnik 2. GMM Workshop*, p. 19.

Lecture Notes on Impedance Spectroscopy – Kanoun (ed)
© 2012 Taylor & Francis Group, London, ISBN 978-0-415-69838-2

Impedance spectroscopy of solar cells

V. Lange, D. Haas, M. Hepting, S. Straub & D. Kühlke
Departement of Computer and Electrical Engineering, Hochschule Furtwangen University, Furtwangen, Germany

ABSTRACT: Impedance spectroscopy is a non-destructive technique with high sensitivity to small changes in the electrical response of a device of interest. This technique has been used to determine the dynamic (AC) characteristics of solar cells. The knowledge of the AC characteristics is of high importance for the efficient action of switching charge controllers. A mismatch can lead to reduced performance of the whole power generating system. The AC characteristics of the solar cell are depending on ambient conditions, in particular on illuminance and temperature. The knowledge of those characteristics can be used to optimize the performance of the complete system.

Keywords: solar cell, impedance spectroscopy, AC characteristics

1 INTRODUCTION

Solar cells are in practice exposed to varying light levels at very different temperatures. These environmental parameters influences the generation and recombination of carriers in the material of any semiconductor device. This affects not only the performance of the solar cell with respect to current generation but also the AC characteristics. The analysis of the AC characteristics of solar cells with impedance spectroscopy (Kumar and Nagaraju 2001) gives valuable information on the parameters of the AC equivalent circuit. Recent work has concentrated on the extraction of basic physical parameters like generation and recombination lifetime (Mora-Sero and Alcubilla 2008) or series resistance and diode factor (Kumar and Singh 2007). Here we focus on the possible adaption to AC applications in connection with switching charge controllers for better performance of the complete power generating system. In a first approach, we have measured and analysed the AC characteristics of a solar cell under different illuminance conditions.

2 EXPERIMENT

Measurements were done with a HP4284 A LCR-meter in the frequency range from $20 Hz$ up to $1 MHz$. The amplitude of the probing AC signal of the LCR-meter was $10 mV$. Keeping this voltage well below the thermal voltage ensures minimal perturbation on the operating conditions of the solar cell. The solar cell was illuminated with a halogen lamp in a dark box in the range from $0 lx$ up to $80000 lx$. The absolute value of the measured impedance $|Z|$ and the phase ϕ are shown in figure 1 and figure 2 respectively.

It is obvious from the measured data, that the light level considerably influences the impedance. The absolute value of the impedance $|Z|$ varies strongly with illuminance at lower frequencies, especially at lower illuminances, while here the phase is less influenced. The presentation of both data in one diagram in form of the Nyquist plot will be given below together with simulated data.

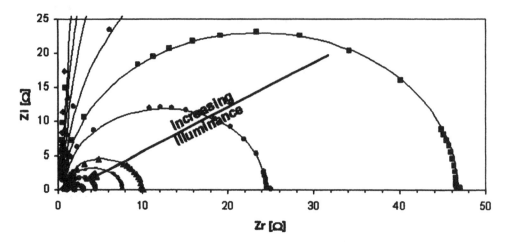

Figure 1.   Measured impedance $|Z|$ as a function of frequency for illuminances between $0lx$ and $80000lx$.

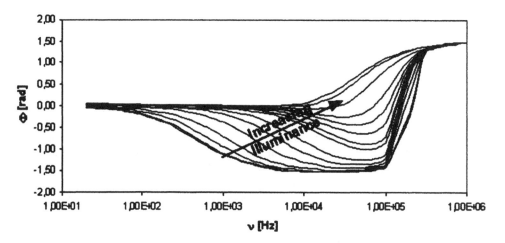

Figure 2.   Measured phase ϕ as a function of frequency for illuminances between $0lx$ and $80000lx$.

3   SIMULATION

The analysis of the data is based on the simplest AC equivalent circuit model of the solar cell consisting of a resistance R_p in parallel with a capacitor C_p, both in series with a further resistance R_s (figure 3).

The complex impedance of this circuit is given by

$$Z(\omega) = Z_r - jZ_i(\omega) \qquad (1)$$

with real and imaginary parts

$$Z_r = R_s + \frac{R_p}{1 + (\omega \cdot R_p \cdot C_p)^2} \quad \text{and} \quad Z_i = R_s + \frac{\omega \cdot C_p \cdot R_p^2}{1 + (\omega \cdot R_p \cdot C_p)^2} \qquad (2)$$

This circuit gives the well known Nyquist diagram given in figure 4.

134

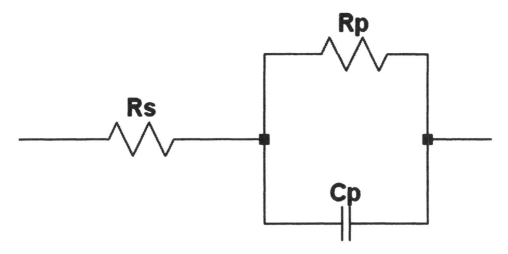

Figure 3. AC equivalent circuit of the solar cell.

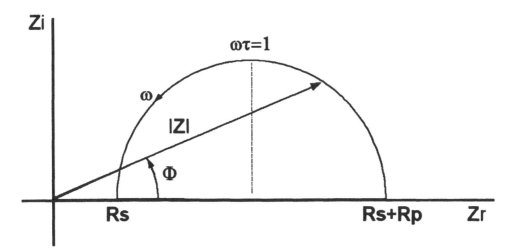

Figure 4. Nyquist diagram of the equivalent AC circuit.

The semicircle with diameter R_p is shifted on the Z_r-axis by R_s and the maximum of the semicircle is related to the capacitance by $\omega_{max} = (R_p \cdot C_p) - 1$. For each measured Nyquist curve, the AC equivalent circuit was simulated with Spice under variation of the circuit parameters R_s, R_p and C_p. Nyquist plots of measured and simulated data are presented in figure 5. Because of the large range of values for Z_r and Z_i, the range of lower values is zoomed out in figure 6.

In the measured frequency range the Nyquist curves are always pure semicircles slightly shifted on the Z_r-axis. Good agreement between measured and simulated data can be achieved. It is obvious from the data, that the solar cell can be fully described by the simple AC equivalent circuit of figure 3. Neither further standard circuit elements nor non standard elements like constant phase element or Warburg impedance (Macdonald 1987) are necessary for the description of the measured data. More complex circuits could lead to ambiguous results for the values of the circuit elements (Macdonald 1987). The semicircles in the Nyquist presentation of the data vary strongly in the lower light level range which corresponds to the strong variations of the absolute value of the impedance at low illuminance

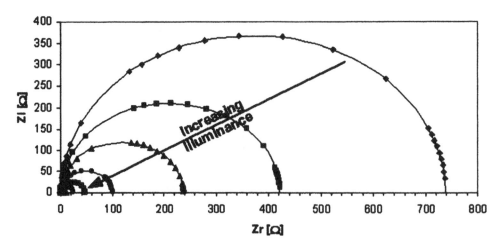

Figure 5. Nyquist plot of measured (dots) and simulated (full line) data.

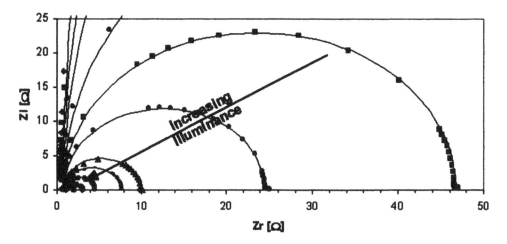

Figure 6. Zoomed Nyquist plot of measured (dots) and simulated (full line) data.

(cf. figure 1). The influence of the illuminance on the parameters of the AC equivalent circuit is as follows: increasing illuminance from $0lx$ up to $80000lx$, which corresponds to nearly full sun light, increases slightly R_s from 0.75Ω to 1.14Ω resulting in an almost constant shift of the semicircle on the Z_r-axis; R_p deceases strongly from 740Ω to 0.85Ω leading to semicircles with shrinking diameter. The shift in the maximal frequency ω_{max} together with the varying value of R_p leads to a strongly increasing value of C_p from $425nF$ up to $5\mu F$.

4 SUMMARY

We have measured the impedance of solar cells at different illuminance. From the analysis of the measured data we can conclude that the impedance Z of solar cells varies with illuminance in particular at low light levels where electrical power is still produced. The analysis of the data has been performed with a basic AC equivalent circuit of the solar cell with only three standard elements, the series resistance R_s, the parallel resistance R_p and the parallel capacitance C_p. Measured and simulated data coincide perfectly. It is not necessary to add further elements like constant phase element or Warburg impedance. The results of our work will allow the adaption of switching charge controller for better performance of solar power devices at lowlight levels. Future work will complete the data to the second environmental parameter, the temperature to allow regulation of solar cell based charging systems to illuminance and temperature.

ACKNOWLEDGEMENT

The work has been financially supported by MPC group Baden Württemberg.

REFERENCES

Kumar, R.A., S.M. and Nagaraju, J. (2001). *Rev. Sci. Instrum. 72*, 3422.
Kumar, S., S.R.C.G. and Singh, P. (2007). *J. Optoelectron. Adv. Mater. 9*, 371.
Macdonald, J. (1987). *Impedance Spectroscopy Emphasizing Solid Materials and Systems*. New York: John Wiley & Sons.
Mora-Sero, I., L.Y.G.-B.G.B.J.M.D.V.C.P.J. and Alcubilla, R. (2008). *Sol. Energy Mater. Sol. Cells 92*, 505.

Lecture Notes on Impedance Spectroscopy – Kanoun (ed)
© 2012 Taylor & Francis Group, London, ISBN 978-0-415-69838-2

Simulation of continuous spectroscopic bioimpedance measurements for impedance cardiography

Mark Ulbrich
Chair for Medical Information Technology (MedIT), RWTH Aachen University of Technology, Aachen, Germany

Jens Mühlsteff
Philips Research Europe, Eindhoven, The Netherlands

Marian Walter & Steffen Leonhardt
Chair for Medical Information Technology (MedIT), RWTH Aachen University of Technology, Aachen, Germany

ABSTRACT: Impedance cardiography (ICG) is a simple and cheap method to acquire hemodynamic parameters. In this work, the potential of ICG to measure these parameters within a whole frequency spectrum has been analyzed using a simplified model of the human-thorax with a high temporal resolution. Therefore, a simulation has been conducted using the finite integration technique (FIT) with a temporal resolution of $103\,Hz$. It has been shown that although neglecting other physiologic signal sources than the aorta, the results obtained from the simulations produce an ICG signal which correlates excellent with measured ICG signals. In addition, real and imaginary part of the impedance show a similar behaviour over time and reveal cole-cole curves known from bioimpedance spectroscopy. What is more, the maximum temporal derivative of the impedance changes with frequency which is one of the reasons for inconsistency between obtained results from different ICG devices.

Keywords: ICG, complex, bioimpedance, simulation, high temporal resolution

1 INTRODUCTION

One of the most common causes of death in Western Europe is chronic heart failure (CHF). Measures for its severity are hemodynamic parameters such as stroke volume (SV) which can be easily and cost-effectively assessed by ICG measurements. Currently, ICG is not commonly used as diagnosis method, because it is not considered to be valid. One reason is the inaccuracy of the technology itself concerning SV calculations. Another possible reason is that processes in the human body during ICG measurements are widely unknown. One way to analyze where the current paths run and which tissue contributes significantly to the measurement result, is to use computer simulations employing FIT.

In the following, the possibility to enhance ICG measurements to a continuous bioimpedance spectroscopy measurement and the advantages or disadvantages of this approach shall be analyzed.

2 BIOIMPEDANCE MEASUREMENTS

For bioimpedance measurements, two outer electrodes are used to inject a small alternating current into the human body and by two inner electrodes the voltage is measured to calculate the complex impedance. If a frequency spectrum between $5\,kHz$ and $1\,MHz$ is used to

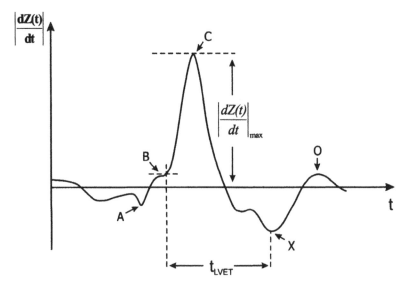

Figure 1. ICG wave with characteristic points.

measure the bioimpedance for each frequency, this method is called bioimpedance spectros-copy (BIS). This frequency range is generally the most interesting one for diagnosis since physiological and pathophysiological processes lead to changes in body impedances with high dynamics. BIS is commonly used to assess the body composition of humans. If only one frequency of this spectrum is used to measure the bioimpedance continuously, this method is called impedance cardiography (ICG). Using ICG, time-dependent hemodynamic param-eters can be extracted from the measured impedance curve (see fig. 1).

The derivation of the impedance changes ΔZ is the ICG signal whose maximum $|dZ/dt|_{max}$ is used for the calculation of stroke volumes. The measured SV by ICG according to Bern-stein and Sramek can be described by the following equation (Water 2003):

$$SV = \delta \cdot \frac{0.17^3}{4.3} \cdot \left| \frac{dZ}{dt} \right|_{max} \cdot \frac{t_e}{Z_0} \tag{1}$$

Here the factor δ is the actual weight divided by the ideal weight, t_e the left ventricular ejection time (LVET) and Z_0 the thoracic base impedance. Since ICG usually operates at a certain frequency between 20 and $100 kHz$, only one continuous point on a complex cole-cole curve can be obtained by ICG measurements.

3 METHODS

Classical ICG analyzes the impedance of the thorax, approximating its volume by one outer cylinder with a conductivity from a mixture of tissues, containing another cylinder repre-senting the aorta. This is of course an assumption which leads to modeling errors. What is more, studies have shown that SV can be measured unreliably only (Cotter 2006). As a result, the task is to find other techniques and models to improve the reliability of ICG measurements. One possibility to do this is to enhance the classical ICG measurements by taking a spectrum of frequencies into account. This shall be simulated using FIT and an anatomical data set of a male human as a basis for a simplified dynamic model. The data set is based on the Visible Human Project Data Set from the National Library of Medicine in Maryland (National Library of Medicine, "The visible human project." [Online], Available: http://www.nlm.nih.gov/research/visible/visible human.html). Since this dataset contains no information about dynamics, a new model had to be created using simple geometries, such as frustums, spheres and cylinders in order to reduce the simulation time. For every expansion step a new model has been created using an aortic diameter increase of 20% as the maximum (Lnne 1992). Since the expansion of the aorta is proportional to the aortic blood pressure (see eq. 2), real measured data from PhysioNet (National Institute of Biomedical Imaging and Bioengineering: PhysioNet—the research resource for complex physiologic signals [Online], Available: http://www.physionet.org/) has been used as basis for the aortic expansion (Greenfield 1962).

$$\Delta R = \frac{\Delta P \cdot R_0 \cdot extensibility}{100} \tag{2}$$

Here R_0 is the diastolic radius of the aorta. The aortic blood pressure has then been scaled to fit the requirement for the maximum aortic expansion so that the diameter of the aorta varies between 25 and $30mm$. For every point in time, the impedance of the whole setup has been calculated so that the impedance depends on the expansion of the aorta only.

The calculation frequency ranges from $5kHz$ to $5MHz$, covering the measuring range of common BIS-analyzers. Conductivity and permittivity values for every tissue in this frequency range have been implemented using the data from Gabriel et al. (Gabriel 1996). The time resolution comprises 103 points in time within a heartbeat. For validation purposes, the simulated curves for $100kHz$ have been compared to real measured data. This data has been acquired using the Niccomo device from medis Germany, Ilmenau.

4 RESULTS

The model created for the simulation is shown in fig. 2. The model consists of all important tissues (muscle, fat, bone, heart, lung, abdominal tissue, important blood vessels) with the aorta as the only dynamic source. The D-Field results of a simulated measurement are presented here by arrows floating through the thorax from abdomen to neck.

ICG signals have been computed at 11 frequencies within the simulated frequency range. The dynamic impedance measured at $100kHz$ has been compared to measured data of a male human and the results show excellent agreement (r = 0.94) using a correction factor. In addition, it is possible to extract a cole-cole curve for every point in time (see fig. 3).

In this plot, the time dependency is not visible for scaling reasons because it is too small compared to the frequency dependency. The magnitude of the real as well as the imaginary part of the impedance change with frequency while keeping their temporal behavior. In addition, the thoracic base impedance Z_0 decreases with increasing frequency due to the cole-cole

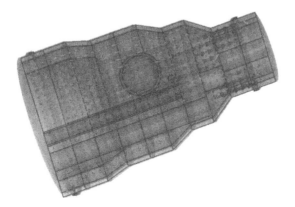

Figure 2. Simulation model and results of a simulation.

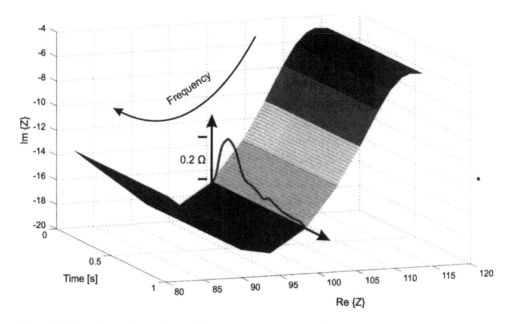

Figure 3.   Time dependent cole-cole plot.

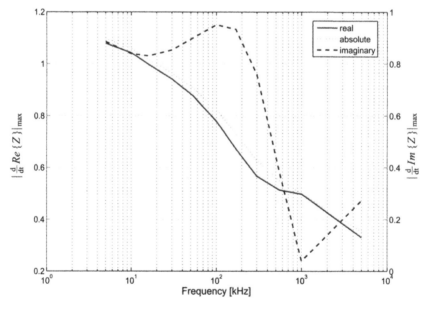

Figure 4.   Frequency dependent behavior of the maximum temporal derivative of the real, absolute and imaginary impedance.

behavior of the complex impedance. Figure 4 shows that the global maximum of the temporal derivative of the real and the imaginary part of the impedance also change with frequency.

This means that $|dZ/dt|_{max}$ of equation (1) is frequency dependent. To sum up, every impedance related variable of this equation shows a frequency dependency within the analyzed frequency range and thus the calculated stroke volume, too.

5 DISCUSSION

The task of this work was to analyze the potential of BIS to improve ICG measurements. Promising results have been obtained using measurements and simulations.

First, simulations with a high temporal resolution have been conducted using the human aorta as the only dynamic source for the ICG signal. In addition, physiological data and a simple model on the basis of anatomical data have been used to produce realistic results. Second, the results of ICG simulations at $100kHz$ show excellent agreement with measured data ($r = 0.94$). Third, the complex impedances have been simulated for multiple frequencies showing cole-cole behavior for each point in time. In addition, the results reveal that complex and imaginary impedances show a similar behavior over time for the whole simulated spectrum and that their maximum temporal derivative is frequency dependent. Thus, it is not astonishing that different ICG devices provide different results when measuring the same patient.

It has been shown that it is possible to measure an ICG signal at various frequencies without loosing information of the aortic signal so that there is no frequency restriction for signal acquisitions. With adapted models or equations for each frequency, this can be used to get additional data about the thoracic composition to detect e.g. edema during their formation phase. Another advantage would be the avoidance of an influence of the orientation of erythrocytes (Visser 1976) on the ICG signal because at high frequencies the membrane of erythrocytes does not longer represent a barrier for the current flowing through the blood.

This model has of course the potential to be improved. One could add more static sources, such as rib cage, kidneys and liver and more dynamic sources should be taken into account including heart beat, lung perfusion and erythrocyte orientation. In addition, the influence of pathologies on the ICG signal could be analyzed.

ACKNOWLEDGMENTS

This work has been supported by Philips Research Europe and contributes to the project "HeartCycle" of the European Union.

REFERENCES

Cotter, G. (2006). Impedance cardiography revisited. *Physiological Measurement vol. 27*, pp. 817–827.

Gabriel, C. (1996). The dielectric properties of biological tissues. *Physics in Medicine and Biology vol. 41*, pp. 2231–2249.

Greenfield, J. (1962). Relation between pressure and diameter in the ascending aorta of man. *Circulation Research vol. 10*, pp. 778–781.

Lnne, T. (1992). Diameter and compliance in the male human abdominal aorta: Influence of age and aortic aneurysm. *European Journal of Vascular Surgery vol. 6*, pp. 178–184.

National institute of biomedical imaging and bioengineering: Physionet—the research resource for complex physiologic signals.

National library of medicine—the visible human project.

Visser, K. (1976). Observations on blood flow related electrical impedance changes in rigid tubes. *European Journal of Physiology vol. 366*, pp. 289–291.

Water, J.V.D. (2003). Impedance cardiography—the next vital sign technology? *Chest vol. 123*, pp. 2028–2033.

Lecture Notes on Impedance Spectroscopy – Kanoun (ed)
© *2012 Taylor & Francis Group, London, ISBN 978-0-415-69838-2*

Tissue diagnostics using inductive impedance spectroscopy

M. Heidary Dastjerdi
*Institute of Measurement Engineering and Sensor Technology, University of Applied Sciences HRW,
Mülheim an der Ruhr, Germany*

C. Sehestedt
Tyco Electronics AMP GmbH, Speyer, Germany

J. Weidenmüller
University of Applied Sciences Koblenz, Remagen, Germany

C. Knopf & J. Himmel
*Institute of Measurement Engineering and Sensor Technology, University of Applied Sciences HRW,
Mülheim an der Ruhr, Germany*

O. Kanoun
Chemnitz University of Technology, Chemnitz, Germany

ABSTRACT: The diagnosis of focal and global tissue structure changes play in both, the transplant surgery and the minimalinvasive surgery, a crucial role. In many application areas of medicine, there is currently no measurement system that can capture these structural changes, but an experience knowledge base of the surgeons and pathological findings, which are time-consuming. In this paper the physical principles of a measuring device that detects structural tissue changes by applying high frequency electromagnetic fields is presented.

Keywords: tissue changes, permittivity, conductivity, impedance, gradiometer

1 INTRODUCTION

In tissue diagnosis so called generalized and focal tissue changes are examined. In order to detect focal tissue changes such as tumors abscesses etc. the area of medical engineering offers Imaging Systems which can determine this kind of tissue change with high spatial resolution. For generalized tissue changes such as fatty degeneration, hepatitis, fibrosis, etc. particularly in the time-critical transplant surgery, there is currently no measurement system but rather the experience knowledge of the surgeons and pathologists. In application areas where time isn't such an important factor pathologic analysis lead to diagnostic results. It would be a great advantage to develop a measuring system, which detects tissue changes. Currently, there are many publications dealing with the dielectric properties, especially the specific conductivity of biological tissue in order to specify tissue changes. (Riedel 2004) However, there is no measuring system, which determines non invasiv, contactless, in vivo, precisely and with high local resolution over a wide frequency range the permittivity and specific conductivity of tissues and thus a fingerprint of the tissue. In this paper a physical measurement principle based on impedance spectroscopy is presented, which is suitable for this application.

2 DIELECTRICAL PROPERTIES OF BIOLOGICAL TISSUE

Biological tissues are differentiated by their dielectrical parameters such as their relative permittivity and specific conductivity. If the medium is considered as a conductor with capacitive properties, it is characterized with the complex conductivity. The complex admittance describes both the complex conductivity and the complex permittivity with:

$$\underline{Y} = G + j\omega C = \left(\frac{A}{d}\right)\left(\sigma' + j\omega\varepsilon_0\varepsilon_r'\right)|Y| = 1S \tag{1}$$

In equation 1 (Grimmes and Martinsen 2008) C describes the capacitance, G the conductance, A the area of plane electrodes, d the electrode separation distance and ω is the angular frequency. The complex conductivity and permittivity and their relation are defined as follows (Riedel 2004; Grimmes and Martinsen 2008):

$$\underline{\sigma} = \sigma' + j\omega\sigma'' \quad [\sigma] = 1\frac{S}{m} \tag{2}$$

$$\underline{\varepsilon} = \varepsilon_r' - j\omega\varepsilon_r'' \quad [\varepsilon] = 1\frac{As}{Vm} \tag{3}$$

$$\underline{\sigma'} = -\omega\varepsilon_0\varepsilon_r'' \tag{4}$$

$$\underline{\sigma''} = \omega\varepsilon_0\varepsilon_r' \tag{5}$$

The real parts σ' of the complex conductivity $\underline{\sigma}$ is related to the current density, and the imaginary part sσ'' to the displacement current density. ε_r is the real part of the relative permittivity ε_r and describes the ability of biological tissue to store the energy of an electric field. The imaginary part ε_r'' describes the loss factor. ε_0 is the permittivity in vacuum. These values are heavily frequency dependent (Sehestedt, Heidary Dastjerdi, Dirsch, Dahmen, and J. 2009; Schwan 1963; Pethig and Kell 1987; Cole and Cole 1941; Foster and Schwan 1989).

In this paper, the characteristics of biological tissue are studied applying impedance spectroscopy. To measure the impedance of an object, electrical energy is fed to the object and the impedance is estimated by using Ohm's law, directly or indirectly.

The complex impedance $\underline{Z}$ is defined as the inverse of the complex admittance:

$$\underline{Z} = \frac{1}{\underline{Y}} = R + jX = \frac{\frac{A}{d}\sigma' - j\frac{A}{d}\omega\varepsilon_0\varepsilon_r'}{\left(\frac{A}{d}\sigma'\right)^2 + \left(\frac{A}{d}\omega\varepsilon_0\varepsilon_r'\right)^2} = \frac{d}{A}\frac{\sigma' - j\omega\varepsilon_0\varepsilon_r'}{(\sigma')^2 + (\omega\varepsilon_0\varepsilon_r')^2} \quad [Z] = 1\Omega \tag{6}$$

The resistance R is the real part of the impedance and the reactance X the imaginary part.

The complex specific impedance z of biological tissue, which is the complex resistivity, is defined as

$$\underline{z} = \frac{1}{\underline{\sigma}} = \frac{\sigma' - j\omega\varepsilon_0\varepsilon_r'}{(\sigma')^2 + (\omega\varepsilon_0\varepsilon_r')^2} \quad [z] = 1\Omega m \tag{7}$$

The electrical conductance of biological tissue, consisting of extracellular fluid and cells containing the intracellular fluid inside the cell membrane, can be determined by its components. Figure 1 shows the simplified equivalent electrical circuit of a cell, which describes the main components of a cell. Compared to the other components of biological tissue the conductivity of the double-layer plasma membrane can be neglected, which leads to a very high value of R_m. The membrane acts, because of its capacitance c_m, as an insulator at low frequencies. In this case the cell can not be penetrated by the current, which means that the current flows around the cell. At very high frequencies the current flows indiscriminately through the extra and intracellular structures. That's the reason why tissue diagnostics are

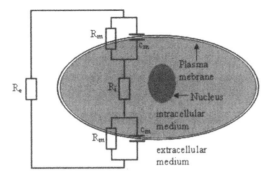

Figure 1. Equivalent electrical circuit of a cell.

applied in frequency ranges from $100kHz$ up to $1MHz$, the β dispersion range. In this dispersion area a part of the current is able to flow through the cell penetrating it and the other part of the current flow around the cell (Fichtner 2002).

3 CONDUCTIVITY MEASUREMENT

3.1 *Physical background*

A contactless method to measure the complex conductivity of biological tissue is an axial gradiometer. The measuring results are usually the corresponding voltages induced in receiver coils, which means that an inverse eddy current problem has to be solved in analytical description.

Axial gradiometers have various application areas and advantages in magnetic induction measurements. It can be used in quality assurances of cold worked metal plates with high conductivities up to $10^{6}S/m$, in diagnostics of biological tissues especially tumor diagnostic or analyzing tissue's health state, which can be used for assessment of grafts typically with very low conductivity in the order of $10^{-2}S/m$ (Himmel, Sehestedt, Heidary Dastjerdi, Weidenmüller, Knopf, and Kanoun 2010).

The experimental setup is figured below (Figure 2). It consists of an excitation coil L_P generating a primary magnetic field. This primary field causes eddy currents in a conductive sample, which produce a secondary magnetic field. In this sensor system there are two detector coils (L_{SU} and L_{SL}), arranged symmetrically, coaxially with respect to the excitation coil and inversely arranged, to measure the secondary magnetic field. This arrangement of the detector coils leads to high cancellation of the primary effect of induced voltages U_{SL} and U_{SU} in each coil.

$$U_{res} = U_{SL} - U_{SU} \quad [U] = 1V \tag{8}$$

The secondary signal, detected by the coils LSU and LSL, can be split into real and imaginary components. The real component is the voltage, which is in phase with the voltage induced by the primary magnetic field. The conductivity of a sample is proportional to the imaginary component of the secondary magnetic field detected by the coils L_{SU} and L_{SL}. Hence by measuring the phase of the receiver coils resulting signal in reference to the excitation signal conclusions about the conductivity of the tissue can be drawn.

The linear combination of current density in the tissue and the resulting magnetic flux $\Delta\phi$ leads to Equation (8).

$$\frac{\Delta\phi}{\phi} \propto -j\omega\sigma \tag{9}$$

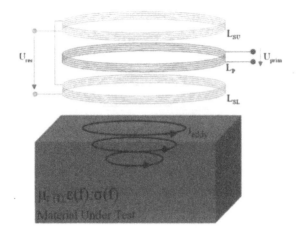

Figure 2. Inductive system.

where $\Delta\phi$ is the induced magnetic flux intensity and ϕ the magnetic flux as excitatory quantity. (Riedel 2004) The phase between the driving voltage U_{prim} and the voltage which is induced by eddy currents U_{res} can be determined by the following equation:

$$\tan(\varphi) = \frac{U_{res}}{U_{prim}} \tag{10}$$

3.2 *Simulation*

The effect, which is used in the measurement of different frequencies is that the electrical properties of tissue (permittivity and conductivity) are frequency dependent. (α, β and γ-dispersion).

A FEA using ANSYS Multiphysics is helpful to get not a quantitative but a qualitative assessment about the eddy current density distribution in the different layers (skin, fat and muscle tissue). The exciting current in the primary coil was set to $I = 100mA$ and a frequency of $\omega = 1MHz$. Figure 3 shows the results. This simulation shows that the largest measurement signal (70%) stems from the muscle layer and the smallest from the skin layer. The spectroscopy, i.e. the measurement with different frequencies, gives thus the ability to move the percentages of the measured signals using the frequency-dependent conductivities.

The thickness of the sample layer and the distance to the gradiometer have a decisive influence on the measurement signal. Thus with the help of the distance and the thickness of the layer a prediction about the origin of individual signal components is possible.

3.3 *Transformer model*

The physical effects, which occur in the experimental setup using a gradiometer, can be illustrated by a transformer model. By calculating the coupling factors within the transformer model, information about the measured parameters can be drawn. The model is influenced by the properties of sensor and material under test. These properties and their affected parameters are listed in Table 1.

The transformer model with resistance R, capacitance C, inductivity L, current I and voltage U is shown in figure 4. The indices give information about the sensor element or the material under test. U stands for upper secondary coil, l for lower secondary coil, p for primary coil (excitation coil) and t for tissue.

148

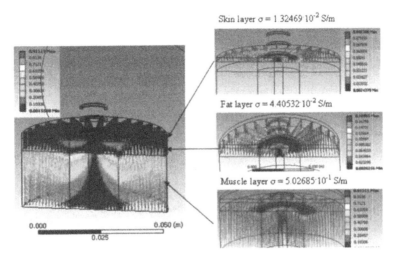

Skin layer σ = 1 32469 10⁻² S/m

Fat layer σ = 4 40532 10⁻² S/m

Muscle layer σ = 5 02685 10⁻¹ S/m

0.000   0.050 (m)

0.025

Figure 3. Eddy current density distribution of a 3-layermodel of biological tissue—simulated in ANSYS.

Table 1. Inter-relation between properties (material under test and sensor) and the affected parameters in the transformer model.

Properties	Affected parameters
Material's conductivity	R_{t1} to R_{tn}, R_{tw1} to R_{twn}
Material's permittivity	c_{t1} to c_{tn}, c_{tw1} to c_{twn}
Material's inductivity	$R_{t1}(T)$ to $R_{tn}(T)$, $R_{tw1}(T)$ to $R_{twn}(T)$
Distance between sensor and material under test	K_{Ct}
Sensor's excitation frequency	w_p
Sensor's conductivity	R_{u1} to R_{un}, R_{p1}, to R_{pn}, R_{l1} to R_{ln}
Sensor's permittivity	c_{u1} to c_{un}, c_{p1} to c_{pn}, c_{l1} to c_{ln}
Sensor's inductivity	L_{u1} to L_{un}, L_{p1} to L_{pn}, L_{l1} to L_{ln}
Sensor's excitation current (non-linearity)	I_p
Sensor's geometric construction	all sensor parameters
Sensor coil's windings	L_{u1} to L_{un}, L_{p1} to L_{pn}, L_{l1} to L_{ln}
(primary and secondary)	R_{u1} to R_{un}, R_{p1} to R_{pn}, R_{l1} to R_{ln}
	c_{u1} to c_{un}, c_{p1} to c_{pn}, c_{l1} to c_{ln}
Temperature of environment	$R(T)$

Taking the mutual inductances M between the coils among themselves and the biological tissue, the coupling factor k can be described with Equations (10) (11) (12), (Riedel 2004; Simonyi 1980).

$$k_{pu} = \sqrt{\frac{M_{pu}M_{up}}{L_p L_u}} = \sqrt{\frac{\phi_{pu}\phi_{up}}{\phi_p \phi_u}} = \sqrt{\frac{\dfrac{\mu I_p(t)}{4\pi}\oint_{s_u}\dfrac{d\vec{s}_p}{|\vec{r}_n - \vec{r}_{sp}|}\dfrac{\mu I_u(t)}{4\pi}\oint_{s_p}\dfrac{d\vec{s}_u}{|\vec{r}_n - \vec{r}_{su}|}}{\dfrac{\mu I_p(t)}{4\pi}\oint_{s_p}\dfrac{d\vec{s}_p}{|\vec{r}_n - \vec{r}_{sp}|}\dfrac{\mu I_u(t)}{4\pi}\oint_{s_u}\dfrac{d\vec{s}_u}{|\vec{r}_n - \vec{r}_{su}|}}}$$

$$= \sqrt{\frac{\oint_{s_u}\dfrac{d\vec{s}_p}{|\vec{r}_n - \vec{r}_{sp}|}\oint_{s_p}\dfrac{d\vec{s}_u}{|\vec{r}_n - \vec{r}_{su}|}}{\oint_{s_p}\dfrac{d\vec{s}_p}{|\vec{r}_n - \vec{r}_{sp}|}\oint_{s_u}\dfrac{d\vec{s}_u}{|\vec{r}_n - \vec{r}_{su}|}}}$$ (11)

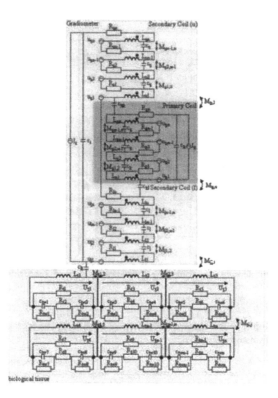

Figure 4.   Transformer model of the measurement setup.

$$k_{pl} = \sqrt{\frac{M_{pl}M_{lp}}{L_pL_l}} = \sqrt{\frac{\phi_{pl}\phi_{lp}}{\phi_p\phi_l}} = \sqrt{\frac{\oint_{s_u}\frac{d\vec{s}_p}{|\vec{r}_n - \vec{r}_{sp}|}\oint_{s_p}\frac{d\vec{s}_l}{|\vec{r}_n - \vec{r}_{sl}|}}{\oint_{s_p}\frac{d\vec{s}_p}{|\vec{r}_n - \vec{r}_{sp}|}\oint_{s_l}\frac{d\vec{s}_u}{|\vec{r}_n - \vec{r}_{s_l}|}}} \qquad (12)$$

$$k_{ct} = \sqrt{\frac{M_{ct}M_{tc}}{L_cL_t}} = \sqrt{\frac{\phi_{ct}\phi_{tc}}{\phi_c\phi_l}} = \sqrt{\frac{\oint_{s_u}\frac{d\vec{s}_p}{|\vec{r}_n - \vec{r}_{sp}|}\oint_{s_l}\frac{d\vec{s}_t}{|\vec{r}_n - \vec{r}_{st}|}}{\oint_{s_p}\frac{d\vec{s}_p}{|\vec{r}_n - \vec{r}_{sp}|}\oint_{s_t}\frac{d\vec{s}_t}{|\vec{r}_n - \vec{r}_{st}|}}} \qquad (13)$$

If the Gradiometer is adjusted, which means if both, the primary and the secondary coils, are equidistant and concentrically arranged to each other, the coupling coefficients k_{pu} and k_{pl} are identical. With increasing distance between the excitation coil C_p and the measuring coil C_u (respectively the excitation coil C_p and C_l), the magnetic fluxes ϕ_{pu} and ϕ_{up} (respectively ϕ_{pl} and ϕ_{lp}),and so the coupling coefficient k_{pu} (respectively k_{pl}) decreases.

Same case occurs changing the distance between the gradiometer and the biological tissue C_t.

3.4   Laboratory setup

Figure 5 illustrates the signal path of the measuring device. The signal generator, controlled through an USB interface, is built up by two direct digital synthesis (DDS) ICs to generate precise frequencies and phase shifts.

150

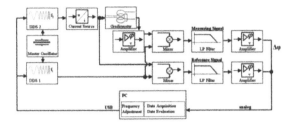

Figure 5.   Signal path of the measuring device.

One DDS generates a mixing signal f_1, which is needed to get a reference signal, and the other DDS generates an input signal f_2 for the current source supplying the excitation signal of the primary coil. The current source is voltage-controlled. The signal of the gradiometers secondary coils is amplified. In order to reduce the amount of equipment, which is necessary for data acquisition, the amplified secondary and primary signals are followed by a frequency down converter (mixer). After a low passfilter and an amplifier the sinusoidal output signals are sampled by a data acquisition card. The software to control the DDS, calculate the phase φ with a least-squares-algorithm and store the sampled data is written in LabVIEW and MATLAB. Trough the described laboratory setup the material under test can be examined with one single frequency as well as with any frequency range given in the user interface.

3.5   *Preliminary investigations and measurement results*

To check the state of the art, first measurement were carried out at a single frequency. For this experimental setup two geometries of gradiometers (Figure 6 and Figure 9) were built in order to detect tissue changes of animal organs.

Both gradiometers are based in a polyoxymethylene cylinder with grooves to guide the copper wires. The current source, preamplifiers and the gradiometer are located in a hand-held unit which is connected to the base station. Signal generation, analogue signal processing and the power supply are located in the base station as well as the electronics for interfacing the computer.

The first gradiometer consists of an exciting coil L_{prim} with 20 turns, which is fed with a sinusoidal current $I_{eff} = 100mA$ at $1MHz$. Concentrically, equidistant with $1mm$ space to the excitation coil the secondary coilswith 10 turns are located. The primary and secondary coils are $10mm$ in diameter. To avoid disturbances caused by electrical interference and mechanical strains the gradiometer is shielded with a $60mm$ grounded aluminum tube section and sealed with potting resin. First the measuring system was investigated using the first gradiometer's geometry (Figure 6) in a clinical trial.[1] In this context, four human resected livers with tumorous and healthy tissue areas were available. Using the gradiometer these two areas were analysed. In literature (Joines, Zhang, Li, and Jirtle 1994) tumorous tissue has a much higher conductivity than healthy tissue of the same tissue type (up two 6 times). Figure 7 shows the measurement results of this first clinical trial. All clinical measurements of human resected livers showed a marked measurement effect between healthy and tumorous liver. These liver resections have different percentages of healthy and tumorous tissue areas. The higher the conductivity the higher the measurement effect, the phase between excitation and measurement coil.

The second gradiometer consists of an exciting coil L_{prim} with 6 turns, which is fed with a sinusoidal current $I_{eff} = 100mA$ at $1MHz$. Concentrically, equidistant with $1.5mm$ space to the excitation coil the secondary coils with 12 turns are located. The coils L are $3.5mm$ in

---

1. In cooperation with the University Hospital of Essen.

Figure 6.   Signal path of the measuring device.

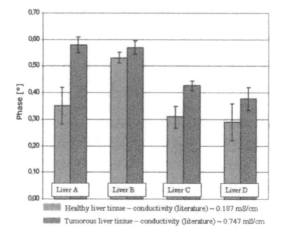

Figure 7.   Measurement results—1. clinical trial.

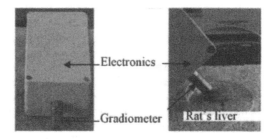

Figure 8.   Gradiometer—2. laboratory setup.

diameter. The gradiometer is shielded with a 12*mm* grounded copper tube section and also sealed with potting resin.

Afterwards the measurement system was investigated using the second gradiometer's geometry (Figure 8) in a clinical trial. In this context, 8 resected rat livers were available. These liver lobes had different fat contents (steatosis) caused by concerted diets of the Lewis rats. Figure 9 shows the measurement results of this second clinical study. The higher the fat content the lower the conductivity of the material and thus the lower the measurement effect.

Clinical measurements of fatty rat livers showed a significant measuring effect, which indicates a high sensitivity of the gradiometer.

152

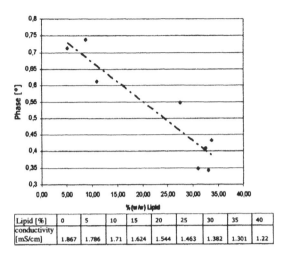

Lipid [%]	0	5	10	15	20	25	30	35	40
conductivity [mS/cm]	1.867	1.786	1.71	1.624	1.544	1.463	1.382	1.301	1.22

Figure 9. Measurement results—2. clinical trial.

4 CONCLUSION

Due to the sensitivity of a gradiometer regarding the conductivity of the material under test, tumorous and healthy liver tissue could be differentiated. Also the grade of steatosis is detectable with a gradiometer.

All measurements were set by a single frequency. In a second step there have to be new clinical studies to investigate different animal and human tissues in a large frequency range ($100 kHz$ to $1 MHz$) in order to get more information aboutthe material under test. In this context comparative measurements with a calibrated laboratory equipment (Lock-In-Amplifier) have to be made to evaluate the developed laboratory setup. Also the coil's geometry has to be optimized for different applications.

REFERENCES

Cole, K.S. and Cole, R.H. (1941). Dispersion and absorption in dielectrics: alternating current characteristics. *Journal of Chemical Physics 9*, 341–351.
Fichtner, W. (2002). *Impedanzmessungen in organischen Fl̈ussigkeiten geringer Leitf̈ahigkeit und ihr Einsatz zur Untersuchung von Schmierölen*. Ph. D. thesis, Dresden University of Technology.
Foster, K. and Schwan, H.P. (1989). Dielectric properties of tissues and biological materials: A critical review. *Biomedical Engineering 17*, 25–104.
Grimmes, S. and Martinsen, R.G. (2008). *Bioimpedance and Bioelectricity*. university of Oslo and Department of biomedical and clinical engineering: Elsevier.
Himmel, J., Sehestedt, C., Heidary Dastjerdi, M., Weidenmüller, J., Knopf, C. and Kanoun, O. (2010). Diversification of the eddy current technology. In *Proceedings of the 7th International Multi-Conference on Systems, Signals and Devices, SSD '10*.
Joines, W.T., Zhang, Y., Li, C. and Jirtle, R.L. (1994). The measured electrical properties of normal and malignant human tissues from 50 to $900 MHz$. *Med. Phys. 21*, 547–550.
Pethig, R. and Kell, D.B. (1987). The passive electrical properties of biological systems: their significance in physiology, biophysics and biotechnology. *Phys. Med. Biol. 32*, 933–970.
Riedel, C. (2004). *Planare induktive Impedanzmessverfahren in der Medizintechnik*. Ph. D. thesis, University of Karlsruhe.
Schwan, H.P. (1963). Determination of biological impedances. *Physical Techniques in Biological Research 6*, 323–407.
Sehestedt, C., Heidary Dastjerdi, M., Dirsch, O., Dahmen, U. and H.J. (2009). Measuring the fat content of liver tissue—experimental setup. In *Proceedings of the 6th International Multi-Conference on Systems, Signals and Devices, SSD '09*.
Simonyi, K. (1980). *Theoretische Elektrotechnik*. VEB Deutscher Verlag der Wissenschaften.

153

Lecture Notes on Impedance Spectroscopy – Kanoun (ed)
© *2012 Taylor & Francis Group, London, ISBN 978-0-415-69838-2*

Impedimetric biosensor for determination of cell viability on cell adhesion

Thomas Frank, Ingo Tobehn, Sabine Nieland & Arndt Steinke
CiS Forschungsinstitut für Mikrosensorik und Photovoltaik GmbH, Erfurt, Germany

ABSTRACT: The term "biosensor" is used when a biological component is directly connected to a signal converter. The sensor captures and transmits information about the physiological changes or the presence of various chemical or biological substances in the environment. Using living cells as sensor elements provides the possibility of a high sensitivity for a wide range of chemically active substances which have influence on the electrochemical activity of the cells. Cell viability is determined by the metabolism or directly through cell division. To evaluate the biocompatibility of active substances or extracts, cell growth can be used. A live-death distinction of biological cells can be determined by the measurement of cell adhesion. The proposed circuit contains an interdigital electrode. The circuit determines the cell adhesion on the change in impedance. The measurement of the impedance is time- and spatial resolution; it shows the growth of cell colonies. This may be a statement about the biocompatibility of the drugs or the extract.

1 INTRODUCTION

The demand for new and effective ingredients for the specific use in therapeutic medicine, the environment or food research are the cause for high R&D activities in the field of biotechnology (Thomas Nacke 2002), (M Brischwein 2005), (Herv Lecoeura and Gougeon 2001), (H. Esfandyarpour 2007). New substances need to be tested for biocompatibility. Various options are available. Cell-based analysis methods are often used. These examine the viability of the cell as a function of the active ingredient, as a function of concentration and time. Two different strategies are pursued, the analysis of metabolic products for the determination of cell vitality and the living-death detection.

2 ANALYSIS OF THE BIOLOGICAL CELL

2.1 *Analysis of the metabolism of the biological cell*

The investigation of the metabolism of the biological cell can be visually achieved, for example, with the help of fluorescent dyes. Nowadays, microsensors are being used increasingly for measurements of the physiological parameters of living cells. The sensor measurement is extra cellular and designed without any impairment or disturbance of cellular life. The range of testable cellular samples is very wide. These microsensors involve pH sensors, pO2 sensors, electrical impedance sensors and sensors for other products of metabolism. For many applications, such as proof of biocompatibility, living-death detection is sufficient. For this application simple and inexpensive measurement techniques and sensors are required.

2.2 *Living-death detection of the biological cell*

In order to realize a living-death detection, a vital or lethal staining is used, for example, the dye trypan blue. The substance is not absorbed by living cells; however, dead cells absorb the dye and become dark blue in colour. A non-inclusion of the dye on the state of the cell membrane is possible. Trypan blue binds to proteins. As trypan blue is toxic, the evaluation must take place immediately after the addition.

Another indicator is triphenyltetrazolium chloride (TTC). The colourless TTC is a redox indicator used to differentiate between metabolically active and inactive tissues. The white compound is enzymatically reduced to red Formazan (TPF) in living tissues due to the activity of various dehydrogenases (enzymes important in oxidation of organic compounds and thus cellular metabolism), while it remains as white TTC in areas of necrosis since these enzymes have been either denatured or degraded.

The death of a biological cell can be detected indirectly by measuring the metabolic products. This requires a measuring chamber for each cell and is not suitable for high-throughput screening. Other indicators are the cell adhesion, the contacts between cells or with extra cellular matrix. Only viable biological cells stick to suitable material. Biological cells and the nutrient solution have different electrical parameters. This allows the detection of the cell adhesion by electrical electrodes. Lethal cells lose adhesion and detach themselves from the electrodes. The information is binary. Through the continuous monitoring of cell adhesion, an evaluation of the colony vitality is possible. The indicator is the rate of cell division. The rate of cell division provides information on the toxicity of the substance or extract used.

The impedance spectroscopy in the range of $1kHz$ and $100kHz$ provides good results referring to the measured cell adhesion. The colony can be connected to the sensor surface by capacitive or resistive coupling. First results show that resistive coupling has a higher sensitivity. A concept for an impedimetric cell sensor is under development.

REFERENCES

Brischwein, M.H. and Grothe, A.O. (2005). Moeglichkeiten und Grenzen der Mikrosensortechnologie in zellulrer Diagnostik und Pharmascreening. *Chemie Ingenieur Technik 12,* 77.
Esfandyarpour, H. and Maiyegun, A., R.D. (2007). 3d modeling of impedance spectroscopy for protein detection in nanoneedle biosensors. *Excerpt from the Proceedings of the COMSOL Conference.*
Herv Lecoeura, Michle Fvrierb, S.G.Y.R. and Gougeon, M.-L. (2001). A novel flow cytometric assay for quantitation and multiparametric characterization of cell-mediated cytotoxicity. *Journal of Immunological Methods Vol. 253, Issues 1–2,* pp. 177–187.
Thomas Nacke, M. and Anhalt, D.F.D.B. (2002). Anwendungsmoeglichkeiten der Impedanzspektroskopie in der Biotechnologie. *tm—Technisches Messen Vol. 69, Issue 1,* pp. 12.

Lecture Notes on Impedance Spectroscopy – Kanoun (ed)
© 2012 Taylor & Francis Group, London, ISBN 978-0-415-69838-2

Quantitative investigation of electrode geometry for biological tissues measurement

Mahdi Guermazi
Chair for Measurement and Sensor Technology, Chemnitz University of Technology, Chemnitz, Germany
Research Unit on Intelligent Control, Design and Optimization of Complex Systems (ICOS),
University of Sfax, Sfax Engineering School, Sfax, Tunisia

U. Troeltzsch & Olfa Kanoun
Chair for Measurement and Sensor Technology, Chemnitz University of Technology, Chemnitz, Germany

ABSTRACT: Impedance based methods are of a big importance for measurements of biological tissues because of the possibility to get detailed information and to realize online measurement systems at low costs. One essential part for this purpose is the study of the effects which are related to electrode geometry considering distance between electrodes and length embedded into the sample. After experimental investigations, a model based evaluation was carried out considering distributed parameter effects. The results show, that deeply inserted electrodes into the sample, which are located close to each other, give the most stable values for model parameters such as the resistive series element (R_E) and the capacitive element (C_{dl}). Geometry changes do not have any influence on the charge transfer resistance.

Keywords: Impedance spectroscopy, contacting effects, electrode geometry, modelling

1 INTRODUCTION

Many laboratory methods can be principally adopted for measurements of biological tissues, such as meat. They require in most cases a complicated measurement setup, and lead generally to high costs (Damez & Clerjon 2008). Therefore they cannot be adopted very often andare generally not accessible for end users. Impedance spectroscopy is a promising method because of the possibility to get detailed information and to realize an online measurement system at low costs.

For measurement of biological tissue, it is important to optimize sample contacting because it influences decisively the quality of measurements. One possibility is to use needles electrodes. These have the advantage to be able to measure deeply within a meat sample and therefore to give an integrative measurement. Therefore they are suitable for experimental investigations for the development of a new measurement method at laboratory stage. Damez et al. used in (Damez, Clerjon, Abouelkaram & Lepetit 2007a) needle electrodes spaced of 15*mm* with a length 5*mm* to study the age of beef. In (Damez, Clerjon, Abouelkaram & Lepetit 2007b) he investigated the dielectric behaviour of beef using an electrode distance of 50*mm* and an electrode length of 5*mm*. Altman et al. (Altman & Pliquett 2005), (Altman, Pliquett, Suess & von Borell 2005) used needle electrodes to predict the intramuscular fat and the carcass composition in lambs of similar weight using the lengths 3, 25 and 35*mm*. Lepetit et al. (Lepetit, Sale & Favier 2002) used two rows of cylindrical needles at a distance of 10*mm* to each other to measure the electrical impedance in bovine meat. García-Breijo et al. (García-Breijo, Barat, Torres, Grau, Gill, Ibánez, Alcaniz, Masot & Fraile 2008) made a measurement of conductivity during the process of meat salting using two electrodes which are placed at a distance 10*mm* to each other.

In most cited literature no explanation was given for the choice of electrode geometry. Only few authors are reporting about theoretical or experimental comparative studies. In this work we will use values for electrode distance a length in the same range. We will vary these parameters and investigate the effect on experimental results. The aim thereby is to find optimal conditions for meat measurements afterwards.

2 EXPERIMENTAL INVESTIGATIONS

For experimental investigation we use agar-agar as material sample. It is a gelatinous substance derived from red algae that is usually used as homologue to biological tissues. In order to emulate different capacitive and resistive behavior of tissues salinity of the phantom was varied in a relative big range using $3g$ and $21g$ of salt. Because temperature influences the conductivity of materials we decides to make the measurements at tow temperatures including typical refrigerator temperature ($T4: T = 4°C$) and room temperature ($T24: T = 24°C$).

Measurements under different conditions are also necessary to verify the choice of electrodes under typical application conditions. Three phantoms composed of $7g$ of agar and $300g$ of distilled water for each phantom are prepared in order to make the measurements in parallel avoiding weight losses of the phantoms.

We used a probe consisting of 2 steel needle electrodes (Diameter of $1mm$) and coated with $1m$ of gold in order to have an inert behaviour. The distance between electrodes was varied at $10mm$, $25mm$ and $50mm$. The length of the inserted part of the electrodes into the agar-agar was varied at $10mm$, $20mm$ and $30mm$.

The measurements were carried out using an impedance analyzer, Agilent 4294 A at frequencies between $40Hz$ to $300kHz$ (figure 1).

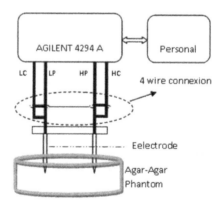

Figure 1.   Measurement setup.

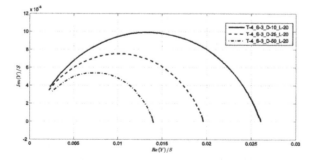

Figure 2.   Nyquist plot in the admittance plane for different distances with length $20mm$ at the temperature ($T4: T = 4°C$) and for phantom prepared with the salinity ($S3: 3g$ of salt).

158

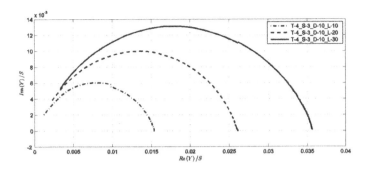

Figure 3. Nyquist plot in the admittance plane for different distances with length 10mm at the temperature ($T4$: $T = 4°C$) and for phantom prepared with the salinity ($S3$: 3g of salt).

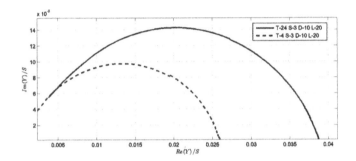

Figure 4. Nyquist plot in the admittance plane at different temperatures ($T4$: $T = 4°C$ and $T24$: $T = 24°C$) for the length 20mm and the distance 10mm and for phantom prepared with the salinity ($S3$: 3g of salt).

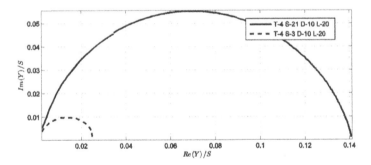

Figure 5. Nyquist plot in the admittance plane phantoms prepared with different salinities ($S3$: 3g of salt, $S21$: 21g of salt) for the length 20mm and the distance 10mm and at the temperature ($T4$: $T = 4°C$).

According to the experimental results, the impedance spectrum maintains a similar behaviour, but the impedance values are significantly changing.

As theoretically can be expected is the impedance becoming higher for a bigger distance and smaller length (figures 2 and 3).

For higher temperature, the impedance decrease (figure 4), which is according to the Arrhenius law.

For higher salinity, the impedance decrease (figure 5). The conductivity increases due to the increase of ion number.

In order to investigate these changes quantitatively we decide to model the impedance spectrum using an equivalent circuit and to make a direct comparison of model parameters for the different cases.

3   RESULTS AND DISCUSSION

3.1   *Mathematical model*

The measurements show a depressed semicircle in the admittance plane (figures 2–5) similar to the equivalent circuit consisting of a combination of a constant phase element (CPE) and two resistors (figure 6).

Data have been fitted to an equivalent circuit or a model equation. For the determination of the model parameters, an evolutionary algorithm described by Kanoun et al. (Tetyuev & Kanoun 2006) and also used by Troeltzsch et al. (Troeltzsch, Kanoun & Traenkler 2006), was applied. Figure 7 shows an example of fitted data and figures 8–11 illustrate the resulting model parameters R_E, R_{ct}, K_a and α for the three phantoms of agar-agar.

The mathematical model is given by equation 1.

$$Y_{model} = \frac{1}{R_E + \frac{1}{k_a(j \cdot 2 \cdot \pi \cdot f)^{\alpha} + \frac{1}{R_{ct}}}} \tag{1}$$

R_E represents the electrolyte resistance; the resistance in any sample is related to the dimensions of the test circuit and the volume resistivity ρ. The expression of R_E for the case of a homogenous field is

$$R_E = \rho\left(\frac{D}{L \cdot d}\right) \tag{2}$$

Where ρ represents the volume resistivity (material property) inherent in the material, D represents the distance between the electrodes, L represents the length of electrode inserted

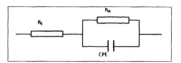

Figure 6.   Equivalent circuit.

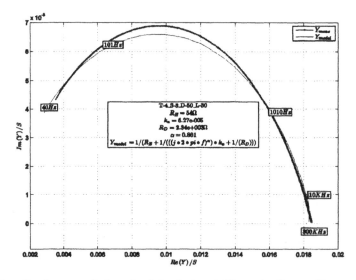

Figure 7.   Data fitted to an equivalent circuit for the length 30*mm*, the distance 10*mm* at the temperature (*T*4: *T* = 4°*C*) and for phantom prepared with 3*g* of salt.

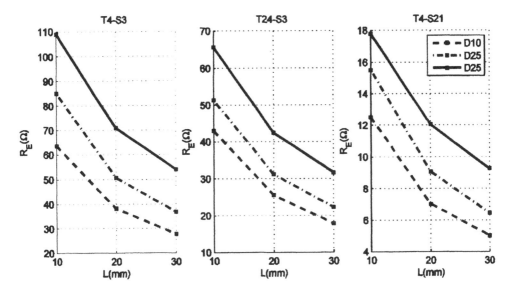

Figure 8. Model parameter R_E for different lengths and distances of the electrode at different temperatures (*T4*: $T = 4°C$ and *T24*: $T = 24°C$) and for phantoms prepared with different salinities (*S3*: 3g of salt, *S21*: 21g of salt).

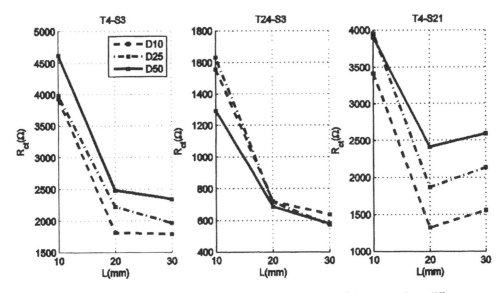

Figure 9. Model parameter R_{ct} for different lengths and distances of the electrode at different temperatures (*T4*: $T = 4°C$ and *T24*: $T = 24°C$) and for phantoms prepared with different salinities (*S3*: 3g of salt, *S21*: 21g of salt).

into the meat and *d* is the diameter of electrode. According to equation 2, the distance between electrodes is directly proportional to the resistance while the length is inversely proportional to the resistance. It was showed that R_E parameter change against geometry agree with theory (figure 8).

R_{ct} represents the transfer resistance. The resistance value is given by the equation 3 (Hamann & Vielstich 2005).

$$R_{ct} = \frac{R \cdot T}{n \cdot F \cdot i_0} \qquad (3)$$

161

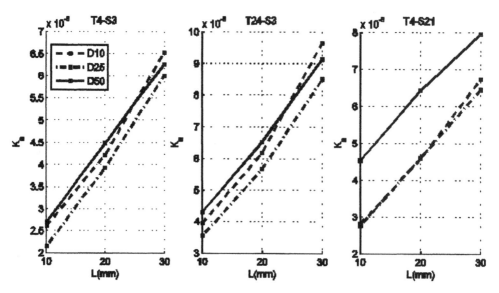

Figure 10. Model parameter K_a for different lengths and distances of the electrode at different temperatures (T4: $T = 4°C$ and T24: $T = 24°C$) and for phantoms prepared with different salinities (S3: 3g of salt, S21: 21g of salt).

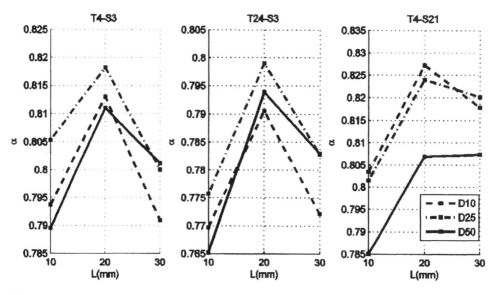

Figure 11. Model parameter α for different lengths and distances of the electrode at different temperatures (T4: $T = 4°C$ and T24: $T = 24°C$) and for phantoms prepared with different salinities (S3: 3g of salt, S21: 21g of salt).

Where T is the temperature, R is the gas constant, n is the number of electrons involved, i_0 is the exchange current density and F is the Faradays constant. The geometrical change does not make any influence on the transfer charge resistance.

CPE can be interpreted as a dispersion capacity or changes of capacity depending on the frequency (Fricke 1932). By investigating the dispersion and adsorption on the dielectric,

Cole and Cole (Cole and Cole 1941) found a behaviour of CPE. The impedance of CPE can be e.g. described according to the following equation of Zoltowski (Zoltowski 1988).

$$Z_{CPE} = \frac{1}{K_a(j \cdot 2 \cdot \pi \cdot f)^\alpha} \qquad (4)$$

When α is close to 0, the Constant Phase Element describes a resistance. Close to -1, it describes an inductance. Close to 1, it describes a capacity and finally, for the value of 0.5, the result is equivalent to the Warburg diffusion impedance.

A conversion of the CPE to a true capacitance according to Hsu et al. (Hsu & Mansfeld 2001) is necessary in order to be able to make a correlation with geometry parameters. Direct physic analyses of the model parameters K_a and α will be not possible withoutconversion (figures 10 and 11).

The CPE conversion to C_{dl} is given by the following equation:

$$C_{dl} = k_a(\omega_m'')^{\alpha-1} \qquad (5)$$

Where $\omega'' = 2 \cdot \pi f$ and f is the frequency at which the imaginary part of the impedance Z'' has a maximum, the value of f is $300kHz$. C_{dl} represents the double layer capacitance at the interface between electrode and electrolyte. It is described in electro chemistry (Bard & Faulkner) by the equation 6.

$$C_{dl} = \left(\frac{2 \cdot z^2 \cdot e^2 \cdot \varepsilon_r \cdot \varepsilon_0 \cdot n^0}{k_B \cdot T} \right)^{1/2} \cdot cosh\left(\frac{z \cdot e \cdot \phi_0}{2 \cdot k_B \cdot T} \right) \cdot A \qquad (6)$$

A is the effective reaction area between the electrode and agar-agar that is equal to the product of the length and the diameter of electrode. k_B is the Boltzman-constant, T is the temperature of the meat sample, e is the charge on the electron, z is the ion charge, n_0 is bulk concentration, ε_r is the dielectric permittivity of meat, ε_0 is the dielectric permittivity vacuum and ϕ_0 is related to the potential at electrode surface. We assume that all the factors are constant except for the dielectric permittivity of meat ε_r.

According to equation 6, area is directly proportional to the length that is directly proportional to the capacitance (figure 12), which is consistent with the experimental measurement (figure 13).

For determining suitable geometrical configuration we calculate the ratios between model parameters for different geometries and compare the resulting rations with theoretically expected ones. For example it expected that at a fixed distance:

$$\frac{R_E(L=10mm)}{R_E(L=30mm)} = 3 \qquad (7)$$

$$\frac{C_{dl}(L=10mm)}{C_{dl}(L=30mm)} = \frac{1}{3} \qquad (8)$$

The results (figure 15) for the parameter R_E ratio at a fixed distance for the three phantoms show the most stable results for R_E at the distances $10mm$ and $25mm$.

For a fixed length the most stable results are observed for the length $30mm$. These results are similar for different temperatures and at different salinity levels.

The results for the parameter C_{dl} show that the distances 10 and 25 give the most stable result at the temperature ($T4: T = 4°C$) and for the phantom prepared with salinity of $21g$ of salt, however for the other situations all the results are very close.

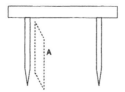

Figure 12. Effective reaction area A.

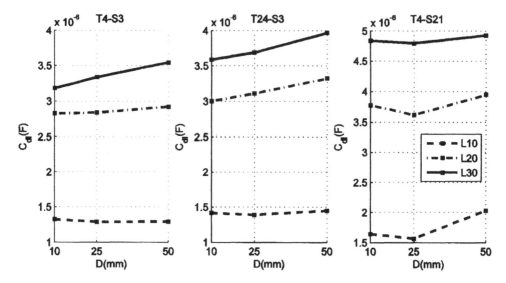

Figure 13. Dependence of the double layer capacitance C_{dl} on the electrode distance and length for different temperatures ($T4$: $T = 4°C$ and $T24$: $T = 24°C$) and for phantoms prepared with different salinities ($S3$: $3g$ of salt and $S21$: $21g$ of salt).

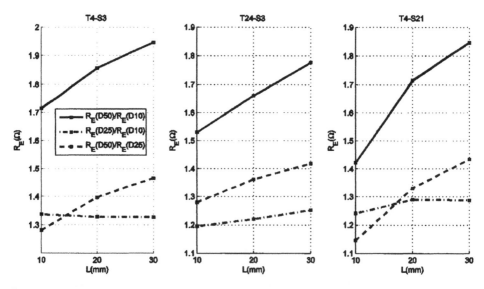

Figure 14. Comparison of parameter R_E ratios for fixed lengths.

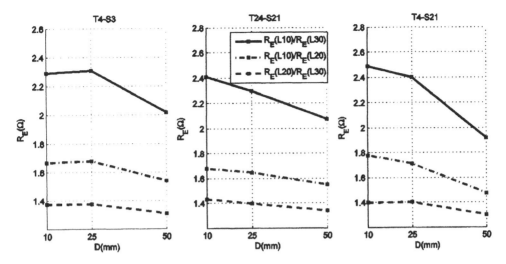

Figure 15. Comparison of parameter R_E ratios for fixed distances.

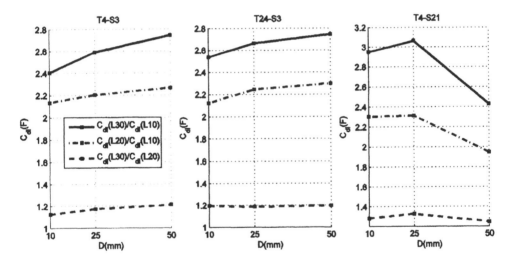

Figure 16. Comparison of parameter C_{dl} ratios for fixed distances.

As a general result we can say that electrodes deeply inserted into the sample and located close to each other show the most stable values for electrolyte resistance (R_E) and for the capacitive element (C_{dl}).

4 CONCLUSION

A theoretical and experimental investigation of the influence of electrode geometry was carried out at agar-agar samples with different salinity and at different temperatures. It was observed that model parameter changes agree with theoretical expectations. Using a model-based evaluation allows a better quantitative comparison of theoretical expectations and experimental results. By comparison of parameter ratios at fixed lengths and fixed distances with theoretical expectations we were able to better identify geometry parameters leading to big deviations.

We conclude that close electrodes deeply inserted into the sample give the most stable values. This investigation gave a first insight on how to find the optimal geometrical electrode configuration for further measurements using meet a sample.

REFERENCES

Altmann, M., Pliquett, U., Suess, R., E.V.B. (2005). Prediction of carcass composition by impedance spectroscopy in lambs of similar weight. *Meat science*, p. 320.

Altmann, M., U.P. (2005). Prediction of intramuscular fat by impedance spectroscopy. *Meat science*, p. 667.

Bard, A., L.F. (2002). *Electrochemical methods fundamentals and applications.* John Wiley & Sons.

Cole, K.S., R.H.C. (1941). Dispersion and absorption in dielectrics: Alternating current characteristics. *J. Chem. Phys. 9(4)*, pp. 341–351.

Damez, J.-L., Clerjon, S., S.A.J.L. (2007a). Beef meat electrical impedance spectroscopy and anisotropy sensing for non-invasive early assessment of meat ageing. *Journal of food engineering*, p. 118.

Damez, J.-L., Clerjon, S., S.A.J.L. (2007b). Dielectric behaviour of beef meat in the 11500kHz range: Simulation with the fricke/colecole model. *Meat science*, p. 514.

Damez, L.-L., Clerjon, S. (2008). Meat quality assessment using biophysical methods related to meat structure. *Meat Science Vol. 80, Nr. 1*, pp. 132–149.

Fricke, H. (1932). *The theory of electrolytic polarization.* Philos. Mag. 14.

Garca-Breijo, J.-M., Barat, O.-L.T.R.G.L.G.J.I.M.A.R.M.R.F. (2008). Development of a puncture electronic device for electrical conductivity measurements throughout meat salting. *Sensors and Actuators A 148*, pp. 63–67.

Hamann, C., W.E.V. (2005). *Elektrochemie*, Volume 4. Edition.

Hsu, C.S., F.M. (2001). Concerning the conversion of the constant phase element parameter yo into a capacitance. *Corrosion 57*, p. 747.

Lepetit, J., Sale, P., R.F.R.D. (2001). Electrical impedance and tenderisation in bovine meat. *Meat science*, pp. 56–57.

Tetyuev, A., O.K. (2006). Method of soil moisture measurement by impedance spectroscopy with soil type recognition for in-situ applications. *Technisches Messen 73*, pp. 404–412.

Troeltzsch, U., Kanoun, O., H.-R.T. (2006). Characterizing aging effects of lithium ion batteries by impedance spectroscopy. *Electrochimica Acta 51*, pp. 1664–1672.

Zoltowski, P. (1998). On the electrical capacitance of interfaces exhibiting constant phase element behaviour. *J. Electroanal. Chem. 443(1)*, pp. 149–154.

Author index

Printed and bound by CPI Group (UK) Ltd, Croydon, CR0 4YY

18/10/2024

01776252-0009